Born in 1959 in Wolverhampton in England. Philip joined the RAF in 1977 as an Aircraft Mechanic—Weapons (AMW or Armourer) and left as a Sergeant after a 22-year career that included tours in Germany and detachments all over the world and saw active duty in the First Gulf War. He used the RAF opportunities for higher education—earning a BSc (Hons) degree with the Open University. Philip also gained degree level Health and Safety qualifications and became a Chartered H&S Consultant. During his second career he held several director level positions working in the construction industry in the UK and Middle East.

When looking at people I need to thank and dedicate this book to, I am at a loss of where to start. The wonderful thing that I discovered as I trawled my memories to write the book was the realisation of just how many people have helped me through life's unpredictable journey. I have named them in the book and recognised their contribution. However, there is one wonderful person who stands tall—my wife, Jules. Without her encouragement there would have been no book, and her professional English language skills helped me to produce all that I could possibly achieve. So, I dedicate this book along with the rest of my life to her.

Philip James

No I'm Not a Pilot

The View From the Ground of an RAF Armourer

Austin Macauley Publishers®
London * Cambridge * New York * Sharjah

All of the events in this memoir are true to the best of author's memory. The views expressed in this memoir are solely those of the author.

A CIP catalogue record for this title is available from the British Library.

ISBN 9781035840748 (Paperback)
ISBN 9781035840755 (Hardback)
ISBN 9781035840762 (ePub e-book)

www.austinmacauley.com

First Published 2024
Austin Macauley Publishers Ltd®
1 Canada Square
Canary Wharf
London
E14 5AA

20241003

I would like to thank the members of the ex-RAF armourer community that I have been able to contact to validate the stories that they were involved in and allowing me to publish them.

Table of Contents

Foreword

I think that before you even start reading this book, I should provide an introduction to set expectations, but also to explain what the book is about and how I have been on a journey writing it. The book mainly covers a twenty-two-year period from 1977 to 1999 that I spent in the Royal Air Force, with occasional explanatory dips into my childhood; and so, by definition, is a historical piece of work. However, it is not a historical record of the RAF or even one of the RAF Aircraft (Weapons) trade or Armourer, as we are known amongst other nicknames; I have left that to people far more knowledgeable than me. This is a very personal account of a quiet and shy teenage boy joining the RAF at seventeen years old and being shaped by that experience over the next two decades to become a person that would have been unrecognisable to me as a sixteen-year-old boy.

To make this task harder, as I am sixty-five years old at the time of writing, I am having to dig out memories from over forty years ago. My memory may play me false with some things, especially names, and I have fought the natural tendency to make things more dramatic than they were. If you are hoping for salacious revelations of affairs and one-night stands when we were away from camp, then you will be disappointed. I am not a gossip by nature and I also think that people in glass houses should not throw stones. Also, there is the forces code of 'What goes on during detachment, stays on detachment' which, I think, like the UK Official Secrets Act has no end validity date. Such things happen in all walks of life and such events add very little to any story.

I have deliberately left out any 'unpleasant episodes' that I think could still hurt or really embarrass people today. However, I have written as honest an account as I can, when, even to me, some things seem far-fetched and unbelievable, I can assure you that as far as I am concerned, I have been faithfully true to what happened; and, where possible, I have validated what I have written with some of the people involved, such as Tim Kay, Colin Bradbury and Billy

Anderson. I have also sought technical clarifications and did some 'fact checking' with a number of people such as Eddy Reilly, Mark Godwin, Daz O'Brien and a few others. I am so grateful for their input as they did correct some things that I would have cheerfully published.

Whilst the book is sequential in terms of its timeline, I wrote each chapter as the memories coalesced into events and I was able to recollect what I was feeling at the time. This was not easy for me as I am not really a reflective person and I tend to look at the future and its possibilities and opportunities rather than the past which, good or bad, is fixed forever. As I did not have a great childhood, I quickly learned how to compartmentalise things and hide bad experiences behind locked mental doors. This book needed me to go deep into my memories and open some of those childhood memory doors that had been firmly shut for most of my life and so that stirred up some strong emotions at times. I have learned that my subconscious is obviously pretty good and gets very active at night, as often I would wake up suddenly in the night with new memories, ideas and thoughts that were new to me. I have added references to UK and international news and political events as little time place markers to give context to what was happening in the wider world.

The obvious question is *why did I decide to write the book?* Well, I was actually going to write about my aunt and grandmother who, in the late 1960s, turned a ruined cottage next to a stream into a home in rural mid-Wales; and, as I was living there for two years, I have vivid childhood recollections of what, for me, was one of the happiest parts of my childhood. After my aunt died of cancer suddenly, aged just forty-two, I recorded hours of conversation with my grandmother onto cassette tapes in the mid-1990s and wrote a very poorly written manuscript of their stories of living out in the wilds with no electricity or running water. However, it was not enough material to produce a book worthy of the name and, so I thought, I could make it a book about parallel lives with my life in the RAF contrasting with their lives in Wales.

As I wrote about my time in the RAF, I soon realised that there would be far more material on that, and it would make the book unbalanced. So, could there be enough material to just write about my life in the RAF? It's not as if I had a stellar RAF career and rose through the ranks to do great things. After I wrote the first chapter about joining the RAF, I became aware of how long it had taken to write just that chapter and the considerable amount of research needed to check things I was writing about. Thank God for Google! So, with considerable trepidation, I asked my smart wife, Jules, with her English PhD and who endlessly reads a vast number of books of all genres, to read my initial efforts and pass verdict. This was mainly so I could stop laying on the settee, dredging up the past from my memory, writing about it and worrying once again about my distinctly average golf handicap, and going back to playing Football Manager 21; in which world, I have personally turned Leeds United into a great Champions League winning team.

To my surprise, and apparently even more to hers, she said, 'Well, who knew you can really write. This is an interesting story and I want to read more!'

So, encouraged by this unexpected praise, and with massive editorial support from Jules that challenged me to write in a way that non-military people can get a peep into my military world, and that those of an RAF or forces background will also immediately identify with. Hopefully, I have been relatively successful in this regard. I must also thank the friends I have made in Al Ain, in the UAE, especially Ian Watson and Darren 'Shally' Shallicker and Mark Grant for doing peer reviews and providing both encouragement and valuable feedback.

The RAF armourers world is quite unique, in that, unless you choose to, you never really depart whilst you are breathing, even long after leaving the RAF. I have met many armourers who left the force over twenty years ago and they still call themselves, with immense pride, an armourer. They are certainly not alone as there are several Facebook groups dedicated to RAF armourers and even the RAF Benevolent Fund Charity as a bespoke armourer section. If you want to get an insight into how the armament world sits within the RAF then I suggest you read a recently published book called 'The RAF's Armourers' by Tony Lamsdale and Phil Appleby (ISBN:1399010336).

However, when I left aged forty in 1999, I decided that I needed a clean break from that military world to totally focus on my new post-forces career. I also have this approach when my bulging waistline leads to me agreeing to embark on yet another of my frequent diets when there can be no sweets or treats in the house, or I will hunt them down like a pig after truffles and gorge on them. Having such a pathetic level of self-discipline means that I must be extreme, and abstinence is the way I achieve this. So, I had never attended any air force or squadron reunions or joined Facebook armourer groups until I retired and started this book; and to be honest, I initially just saw Facebook RAF and armourer groups as valuable research material, but that has changed drastically now, and I h a reunion event and took part in the 2024 'Cenotaph Bimble' Remembrance Day.

The RAF not only gave me twenty-two years of very interesting life but it also paid 100% of the costs for me to gain a BSc degree with the Open University. That degree qualification opened doors that would never have been available to me, and, in time with further NVQs, I became a chartered health and safety practitioner with my focus on construction. After two decades in the merciless civilian commercial world, I have had the highs of promotions to board level positions and been head-hunted for amazing construction projects, but also the lows of being let go when things go wrong. I have been fortunate to gain sufficient financial reward and had some personal achievements in the safety world of major construction projects in Abu Dhabi in the UAE that I am very proud of.

Through my book research, I now see people that I knew from my time in the RAF still actively supporting each other in the Facebook ex-armourer community, raising money for charities and other good causes; keeping in touch; having regular reunions, such as the annual celebration of Saint Barbara, who is the patron saint of armourers; the fund-raising events; and reading the moving epitaphs when someone has sadly passed away. This is a great level of armourer camaraderie that clearly not only still exists but is perpetuated by new still serving RAF armourers joining the groups. The only depressing revelation has been on an armourer site that is dedicated to the guys that passed away. There is a constantly updated spreadsheet of names, and it is a shock to see people I once knew that have now been lost to us, and the realisation that one day my name will be on that spreadsheet (but luckily, I will not be able to read it).

However, I realise now how wrong I was to turn away from such a wonderful, unique and caring breed of people; so, once again, with a slightly guilty conscience regarding my abstinence from the armament world, I am now proud to once again call myself an ex-armourer. I hope that my fellow 'plumbers' agree that this book does justice to giving myself such an accolade.

1. Where Do I Sign to Join Up?

In January 1977, as Jimmy Carter was preparing to become the 39th President of the United States, at the same time, a rather weedy, spotty, quite timid and nervous seventeen-year-old boy had said goodbye to his father on a rain-swept and blustery railway platform before boarding a train in Torquay, South Devon, heading to a destination in northern England that I had never heard of. This was not like being taken to university by your parents in a car packed with duvets and creature comforts from home, as we had been instructed to just bring a small holdall with minimum clothes. After six hours and three train changes enroute, I arrived at Newark on Trent in Lincolnshire, relieved that I had completed the journey.

Newark is a small town that serves as a transportation crossroads. Situated on the Trent River, it is where the north-south and east-west railway lines cross, along with the Great North Road (now known as the A1) and the Roman Fosse Road (now called the A46). I left the train and nipped to the station toilet for a nervous pee and then went outside Newark Castle station into the cold winter's day to see a grey RAF bus waiting to take us to RAF Swinderby. My feeling of relief about my safe arrival was now changing into one of slight trepidation. Two other young men, with fashionably long hair, were stood outside the bus smoking cigarettes, but they stubbed them out on the wet ground with their shoe as I approached. They had all been waiting for me, apparently. I boarded the bus and a Corporal in uniform asked me my name and checked it against a list on his clipboard; satisfied, he nodded and told me to take a seat.

The bus already had about twenty guys on it, and some said hello with short nods. I took a spare seat halfway down the bus. The front door swished as it closed with a clunk, and, for a second, I had a feeling of being trapped as the engine, which had been ticking over to keep the heaters blowing, now roared into life as we set off. I looked at my fellow passengers all setting out on the same journey to what seemed would be a very different world; some were just sat quietly, looking out of the steamed-up windows at the bleak winter countryside, others seemed to have made friends during their train journey and so were chatting away, but for most, the conversation was a bit stilted, and any laughter had a nervous ring to it. The air smelt of young men with their stale cigarette breath and cheap aftershave. For the first time, I looked at the person next to me and we started chatting in that sort of non-committal 'nice day' way. But, talking with him settled my nerves and soon I was getting excited again about what lay ahead. It now suddenly felt like a big day, like your first day at school, as we all knew that this was a big deal that was going to change our lives forever.

I have thought long and hard about why people join the Royal Air Force in the first place. You will certainly not get rich or famous whilst in uniform, serving your country, and you have no control over your job or where you will be sent to work, and you may get killed or maimed in a future conflict in a foreign land. Luckily, only a tiny percentage of the population needs to think that this career choice makes sense in order to populate the RAF with young people, unless a major conflict happens, such as the invasion of Ukraine by Russia in 2022, when free-world governments suddenly begrudged defence spending far less. As Putin launched his ill-advised 'special military operation', there were 33,320 members of the RAF. In 2019, before the Covid-19 pandemic, there were around 75,000 people working across 400 different companies at London's Heathrow Airport; so, twice the number of people serving in the RAF.

So, why the RAF? Well, for me it was quite simple—it was an escape from a very troubled home life and a very potent desire for adventure. Even from a young age, I got bored easily, and I still do; I hated routine and loved exploring. When I was four years old, living near Wolverhampton, I went 'missing' on the estate and was found nearly a mile away, driving my little red peddle car. Even before I was ten, I would ride for miles (thirty miles) on my child's bicycle, which had little fat wheels, to Bridgenorth and back; and, in my teens, go twice that far on my racing bike to Shrewsbury, and dream of carrying on into that foreign country called Wales before heading east and home again.

This desire for adventure was all about travel, seeing different cultures, historic places and meeting different people. It was not about needing an adrenalin rush. I have never had the urge to jump out of a serviceable plane parachuting, bungee jump or even hurl myself down the side of a snow-covered mountain on two planks of carbon fibre! Nowadays, I satisfy the same urge to wander with my wife, Jules, in a campervan around Europe and cruises as often as we can. So, the RAF 'sell' of travel and adventure really appealed to me.

I briefly considered the navy and being a submariner, but I liked the outdoors and sports too much, and the army just did not have the same draw as they had no aircraft. Most teenage boys at that time probably had posters on their bedroom walls of supercars, sport stars, and maybe pop stars and a few magazine pictures of scantily clad or naked women hidden from your mum under the bed. I had a poster of Leeds United, who were a big deal in the early 1970s, but of far more importance to me was the large colour picture that dominated my wall of a Supermarine Spitfire Mk5 flying over southern England. I can still see that picture in my mind's eye. It must have been taken in the autumn, due to the variety of shades of green in the lower background, a beautiful backdrop to what must be one of the most beautiful pieces of engineering ever created. Of course, like most teenage boys, I was also fascinated by the magazines about the unclad opposite sex, hidden under my bed!

Whenever I see Vietnam war films in which young men enlist to the armed forces and arrive at their basic training location, or 'Boot camp' as the Americans aptly call it, I feel a certain empathy. Not that in 1970s, the six weeks' basic training course was comparable to the harshness and potentially abusive methods seen in some films depicting US Marine training courses, where bullying, harassment and some unpleasant techniques to try and weed out the quitters and those who are unsuitable early on seem to be employed—methods that I doubt would be acceptable today. RAF Swinderby opened in 1940 as one of the last, hastily constructed bases in the late 1930s, 'Expansion scheme' as home to several bomber squadrons in response to the threat of Germany. In 1964, it became home to No. 7 School of Recruit Training, responsible for the initial military training (known as Basic Training) for all male enlisted personnel.

On arrival at RAF Swinderby, the bus went quiet, and we stepped out into the cold and were made to 'form-up' in ranks of three lines outside the bus. The 'training' started immediately with our motley ranks being shouted at by the Drill Instructors corporals and their sergeant boss, who informed us that we were indeed lowlife scum, and that Her Majesty's Royal Air Force had kindly deemed to make us all into worthy human beings that would be willing to die for their Queen and country. Then, the well-worn arrival process, honed by military precision timing since the camp had become the RAF's basic training camp for airman thirteen years earlier, kicked in. With us being loosely marched (perhaps ‘herded’ is a better adjective) around the camp, carrying an ever-increasing amount of ‘gear’ in the kit bags we were given. Uniforms, gym wear consisting of ill-fitting shorts, white and blue tee-shirts, black socks and white plimsolls, followed by getting our mess tins, cutlery, towels, and later, bedding.

As I had been in the Air Training Corps (or ATC as it’s known), which is an RAF organisation for thirteen to eighteen-year-olds to encourage them to join the RAF, for several years and had been on six weeklong summer camps at various RAF stations, this experience did not phase me too much. Also, for me, this all meant that it was really happening and I was in the RAF, which was something I had wanted to do for years; so, I was actually quite excited and lapping it all up. We were then lined up in a room and had our names shouted out with a three-digit number, which was the last three digits of your new RAF service number.

When I heard ‘James 555’, my first thought was, *great, that is easy to remember*, as I sometimes struggle with numbers. I needn’t have worried because, as every ex-service person will tell you, you will be able to recite the number for the rest of your life; and, despite being out of the RAF for nearly quarter of a century now, for me, F8131555 is still ever present in my mind. I’m confident that even my ex-wife could still recite it backwards! There was some laughter when one of the guys, who was called Bond, was given a last three of 007—so, apparently, some of the ‘pen pusher’ RAF administration clerks do have a sense of humour. We had to sign some paperwork, maybe the Official Secrets Act, and probably a contract; I can’t remember. And then, I was given my first week’s ‘pay’ in advance, in cash.

Walking into the next room, some of that cash was immediately taken from us to pay for a haircut. This was the 1970s, when long hair for men was in fashion and the floor around the barber’s chair was piling up with locks as each recruit was given a number two buzz cut all over in about thirty seconds before being marched out. What a great little number this was for the local village barber, being paid for doing the simplest haircut on the planet to sixty new customers every few weeks. The look of horror on some of the guys’ faces as this happened made me smile at the time, as they went from looking super trendy to looking like a prison inmate in the blink of an eye. When we left that building, we had money in our pockets; an employment contract for three or six years, subject to passing basic and trade training; and very chilly heads in the biting cold January wind.

Next, we were loosely marched to one of the several H-shaped barrack blocks, the number Block No. 6 is in my mind, but I can't be sure. There, we were informed about which squadron we were in and within that which 'Flight' we were assigned to. We then carried our kit and were shown into a large eighteen-man room and allocated a bed, single locker and a small bedside and table. I remember the room was quite long with lots of windows and nine bedspaces down each side; the floor was a shiny linoleum reddish brown surface that we would all soon become very intimate with. I have seen films of criminals arriving at prison and this did feel a little bit menacing like that.

We were shown by the DIs how to make our beds and unmake them into a 'bed pack'—which was basically the folded sheets and blankets made into perfect rectangles and wrapped in a creaseless bed top cover blanket. This would have to be done to perfection each morning or a punishment for the owner of a shit bed pack would ensue. We soon learned to insert hidden pieces of cardboard to get and maintain crisp angles, which, of course the DIs knew we were doing but let us believe we were being innovative and clever! The actual clever guys made their bed pack once and slept in a sleeping bag. We were also shown how to store our kit and clothes properly and then taken to the airman's mess for dinner for some distinctly average food, which was something that the Swinderby Airman's Mess became infamous for.

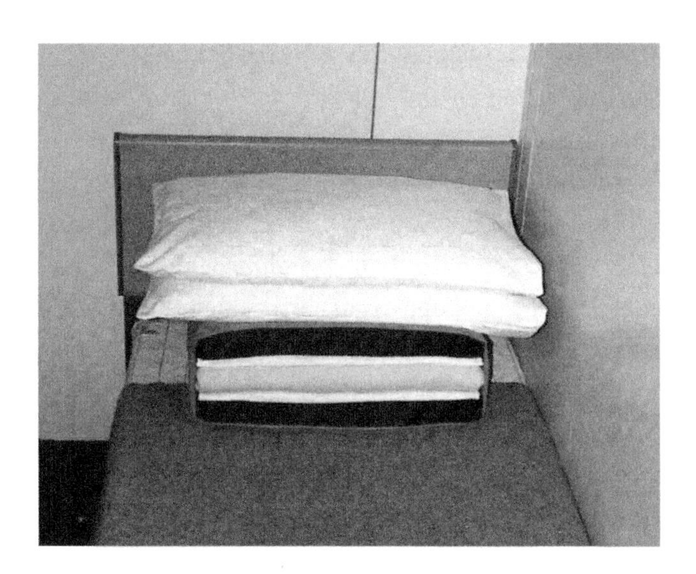

Fig 1. The infamous 'Bed pack' (Courtesy Mark Bean)

We were also shown where the Newcomers Club, the NAAFI bar and entertainment venue was and that concluded our first day in the RAF. Back in the big barrack room that I now shared with seventeen other guys, it struck me that these were young lads from all over the UK. I remember being awestruck at two Geordies rapidly talking to each other and not understanding a word of what they were saying and then a lad from Northern Ireland that I found incomprehensible!

Then, there was a Scottish lad who became Jock; two Welsh guys, one became Taff and strangely, the other, just Colin; a lad with an incredibly broad west country accent, who was instantly nicknamed 'Scrumpy' by the DI's; and Liverpudlian, with a sing-song voice, who was now known as 'Scouse', obviously; finally, there was a deep voiced and very tall cockney from east London; So, quite an eclectic mix. That night, I lay in bed and reflected on my first really long day, which had been a whirlwind of a day and an assault on the senses, but I was content. I was not so happy about the sleeping conditions though, having eighteen young adult males sleeping in the same room does tend to create its own atmosphere of sweat, bad breath and flatulence, all made worse by it being winter and the heating system making the huge iron radiators create a stifling temperature offset by the cold blast of air through the much-needed open windows.

Eventually, as things settled down with the inevitable joker shouting out "Night John Boy" from the Walton's TV series, the room then went quiet, before a variety of snoring started up and I thought I could hear gentle crying coming from somewhere. It seemed like I had just nodded off when the tannoy speaker in the room blasted out reveille at 06.00hrs and all the room ceiling lights came on to blind us. Just to make sure we were on message that it was time to get up, our Corporal DI walked down the room, banging the metal frame of our beds to add to the bedlam. Somehow, one of the guys, called Jacko, still stayed asleep or, not unreasonably, just decided it was too early to get up, and hid under his covers. On spotting this, the Corporal went over to his bed and sarcastically apologised for disturbing him, before tipping the entire bed on to its side and ejecting the now startled and confused occupant, desperately rubbing his eyes to try and see who was assaulting him. We were later told that as Airman Jackson had struggled to get up at 06.00hrs, then to give him more time to gently come around, reveille the next day would be at 05.30hrs, and that perhaps someone could kindly make him a nice cup of tea and bring him a daily newspaper.

"Which paper do you read, Airman Jackson?"

Jackson wisely kept mum and looked sheepish. We all soon learned to leap out of bed when the tannoy sounded, and after a week or so, reveille went back to 6.00am and the Corporal stopped his early morning visits—his work was done!

Of course, we all soon adjusted to hearing the different accents, and friendships started forming straight away. The geographical mix became very apparent on a Saturday afternoon when the football was on the radio and each latest goal resulted in a mix of cheers and groans. After a couple of days, we were told to select our own 'Senior Man' who would oversee the flight and the person all communications went through. As a 'reward' for this additional duty, the Senior Man would occupy the single bedroom at the entrance to the barrack room. The choice was easy and unanimous—Geordie, who was a twenty-seven-year-old that had transferred from the army to join the RAF (we never did find out why); but, obviously, it was a wise choice. Being twenty-seven meant that he was so much older and more mature than the rest of us. It turned out that we made a great choice too as he did a great job and he also mentored quite a few of the lads if they struggled with something in the training, or just got homesick.

A few years later, in the popular TV series Auf Wiedersehen Pet, Dennis Patterson became the de facto leader. I was not surprised as he was just like our Geordie. I realise now that I never ever even knew what his real Christian name was; he was just Geordie, and like a great older brother to us all. I don't know if the RAF mixed up new recruits geographically by design to make everyone adapt to different nationalities, cultures and backgrounds and force us to adapt to each other; I suspect not, but anyway, it worked well.

The British army has a very different approach, having created regiments in geographical locations that new entrants join and often stay with for their entire career. These regiments can move lock, stock and barrel to a new location, taking every single person and their families with them. This has benefits in terms of supporting each other like a huge family, but it also means that you never escape those same people if things like the inevitable affairs, marriage break-ups, and cliques forming happen. In the RAF, you were just a body with a unique number and were 'posted' like a human parcel to wherever there was a space that your training and experience could fill.

Personally, I liked it that every posting meant a fresh start with new people and, often a new job role; though, of course, over time you would always find people you knew from previous postings. The purpose of the six-week basic training course is to prepare you for a service life in the RAF, so they get you super fit, teach you basic field craft, how to fire a gun, make you understand the history and cultural expectations of being in the RAF, and, most importantly, how to be a team member and get along with others. It was also to weed out people that are just not cut out for a service life early in the process. The saying "If you can't take a joke then you shouldn't have joined up" is a humorous take on a very real expectation.

As in all armed forces, the RAF expects you to do things without too much questioning, and 'real-time' situations mean that things do go wrong, and even the best plans can go awry very quickly. So, you do need to have a personality that can accept things not being great, but still effectively do what is required of you. This takes a certain type of personality and it's not for everyone; some people did leave the flight after a week and I felt sorry for them missing out on what I was sure was going to be a great career.

The military also likes to dehumanise each person to an extent so that the training and doctrine take root and orders are obeyed. Making everyone have short hair, so no individual hairstyles; and then covering the head with hats and berets, which hides hair colour; together, with the same-coloured uniform, makes everyone identical except for shape and size and, of course, rank badges that soon become very significant in their importance. Perhaps that is why even nicknames are standardised.

As well as geographical references, such as Scouse, Taffy, Jock, Paddy, Geordie, etc. If you are very short in stature, you may be called 'Lofty'; and, if you are very tall, 'Tiny'. A play on names often happens, so Smith becomes 'Smiffy' and Woods is 'Woody'; anyone with the surname Miller is 'Dusty'. If your name is White, you are known as 'Chalky'. The other way a nickname is assigned is if you are unfortunate to do something that is embarrassing. One of the lads on another flight had diarrhoea and messed his pants during a cross-country run and so he became known as Crapper from that day onwards. When I bumped into him again, many years later, it was still his moniker and probably stayed with him for the rest of his RAF career!

This indoctrination of 'team before self every time' starts on the very first day with the DI's telling you that you are a worthless piece of shit and that they can't understand how you were ever recruited,

"Did your mother send you here for a dare?"

Their aim is to make you dislike them, hate them even, and make it an 'us' versus 'them' situation. The verbal abuse, which is probably what it was, came in the form of military humour, which we soon developed an appreciation for and which most of us would become exponents of for the rest of our lives. Our early efforts at marching, in which every flight would have one guy that 'tick-tock' marched with both the arm and leg on each side of the body swinging at the same time instead of alternatively. This was amusing to see, and caused chaos on the marching ranks; it was usually sorted out very quickly by the DI. Soon, we were marching everywhere to the sound of "left, left, left, right, left" to keep our timing on beat.

We were advised that during marching, we could "Open your legs, your balls won't drop off."

One of the guys made the near fatal mistake of taking a short cut across a grassed area and was advised in a loud screaming voice "Get of the grass! Only God and the Station Warrant Officer (SWO) are allowed on the grass, and God has to ask permission from the SWO to do so!"

All of the DIs had their favourite saying. Our DI liked to say that we could march to wherever we were due next for training, to save us walking, and "Boots will march!"

The harassment and bullying, such as ripping apart your locker because of a speck of fluff on the floor or throwing your bed pack out of the window, is all designed to foster this. Personal hygiene and tidiness are embedded as you had to prepare your bedspace for daily inspection; everything in your locker had to be stored to a predefined layout, and your sheets and blankets were folded into that perfect 'bed pack'.

The DI staff were understanding of our painful transition and demonstrated this through comforting guidance, such as "Do you feel sorry for yourself? Well, sympathy is somewhere between shit and syphilis in the dictionary."

The training weeks passed by with a morning routine of long cross-country runs extended to ten miles according to the accounts of others. Despite it being January, these were done just wearing shorts, tee-shirts, black socks and white plimsolls that, regardless of how wet and muddy they got, were expected to be dried and pristine white again by the following morning. This was an almost impossible task, as the DIs would have known. We were lucky it was winter, and the heating was on full blast; the radiators and boiler room would all be plastered each night with washed plimsolls with the laces removed, trying to dry them out.

After breakfast, you would use the tubes of whitener purchased from the NAAFI shop to get them from grey back to pristine white. I put on damp plimsolls for Physical Training (PT) or the next cross-country run and they would still be damp inside with still drying whitener on the outside. As you ran through the first muddy puddle, you knew all your hard work cleaning your plimsolls had just been destroyed. Of course, the RAF could easily have given us two pairs so we could alternate between pairs, but what would be the fun in that?

There was an awful lot of drill training with old Drill Practice (or DP) .303" rifles that had been decommissioned; and, of course, we marched everywhere, which as I said 'saved you walking', according to the ever-humorous DI's. We also spent hours on the parade square learning marching drill formations, known as 'square bashing' and even marching in sideways formation, which is a useful drill move only performed by the RAF; it led to the army and navy calling RAF personnel 'crabs' as they only move sideways. If the other services are flying on RAF transport aircraft, they call it 'Crab Air'. The rest periods from physical activity were theory lessons on the RAF workings and how we would fit into this new world that we were being shaped for and marched into. I can remember one session on pensions and how the RAF system worked, which at seventeen years old being told about what would happen when you were fifty-five was almost surreal.

We were also taken to the medical centre, lined up and given several injections in both arms at the same time. One of the inoculations was for yellow fever and that evening a few of the guys reacted quite badly to the cocktail of drugs. I remember one guy collapsing whilst he was ironing his uniform trousers. If there had been no cross-country run in the morning, then there was gym or sports activity in the afternoon instead. Here, we met another specimen of sadist—the RAF Physical Training Instructor (or PTI), as they were known.

On finishing their trade training, these guys were given the rank of acting Corporal so they had some rank status and could tell other people (well, lower ranking airmen like us) what to do. Every day, your body was pushed physically. Soon, our bodies were changing, becoming muscled and lean and even the guys that were a bit overweight at the start were now fine athletic specimens. When I joined the RAF, I was around five foot and five inches (163cms) tall and weighed nine stone (57kgs), when I left basic training just a few weeks later, I was 2.5 inches (6.3cms) taller and I weighed 10.5 stone (67kgs). Naturally, after a few weeks, our head hair had actually grown a couple of centimetres, but there were no more mass haircutting sessions.

This abomination by nature was met by the DIs with them saying things like "Do you have a headache airman? Because I'm standing on your bleeding hair—get it cut!"

The most exciting part of the training week was learning to shoot the 7.62mm Self-Loading Rifle (SLR). We had to be able to strip down and reassemble the rifle into its component parts blindfolded and within a deadline. In the classroom, we were taught firing drills and basic marksmanship skills until the RAF Regiment instructors, who were known as Rocks or Rock apes, were satisfied that we could be unleashed safely onto the firing range. During my time in the ATC, I had fired the Le Enfield .303" rifles both in the .22 version at the squadron indoor range and the .303" during ATC summer camps at various RAF stations. Each time, I had the bruises from the recoil kick back on my shoulder to prove it. So, again, it was familiar to me, but the range was a noisy place with the crack of rifle shots and the DI's shouting out fire command orders.

The noise, together with the strong smell of cordite in the air, added to the tension of firing live ammunition. Some of the guys really struggled to even hit the target at first. After one poor guy produced yet another insipid demonstration of shooting skills, the DI provided useful feedback and encouragement with very enthusiastic language and kicked the boots of the now visibly shaking trainee airman. But over time, we all got better and eventually passed the shooting proficiency test so we would not fail basic training. I, along with a few others, managed to shoot well enough to be classed as RAF marksmen and we even got a 'crossed rifles' badge on the sleeve of our uniform. The saying of one of the Rocks that is still with me forty-six years later is "Oh my God, why should Britain tremble with you lot defending it?". To be fair, he probably had a point!

The biggest lesson that we all learned quickly was that when someone 'f*cks-up', the entire room or flight gets punished, not just the individual. You soon realise the need to be one team and that you will suffer even if you are personally one hundred percent and a roommate messes up.

The punishment was often press-ups or an extra drill session on the parade square, but the worst words you could hear would be "Right as a result of airman X not doing what he was supposed to, tonight will be a bull night."

Deep joy for us all and naturally airman X did not have an enjoyable evening and would definitely be on ablutions cleaning duty, absolutely no question! Each week, there was a 'bull night' anyway, when we must all clean the entire barrack block and bring the floors to a shiny finish using a duster under a weighted block on a broom stick called a 'bumper', which we used to remove the grungy and strange smelling orange polish that had been applied to the linoleum floor.

Brasso was used to cleaning the window handles and latches to a shine and the toilets and sinks cleaned and scrubbed. Once the ablutions had been cleaned, to keep them that way, only one shower, sink and toilet were left for use by everyone that night and hastily cleaned in the morning by whoever was lucky enough to be on 'ablutions' cleaning duty. Woe betides anyone that decided they needed to take 'a dump' in any of the other toilet cubicles before the inspection!

Then, at 8:30am came the morning inspection, as reveille had blared over the room speakers at 6:00am. We were already over two hours into our day. Our trousers were pressed with razor sharp creases, shoes polished and toe caps 'bulled' to a high shine as we stood ramrod straight to attention at the bottom of our beds. The DI Corporal, or worse, the DI Sergeant, would enter the silent room and slowly inspect each man and his bedspace moving up one side of the room and then back down the other.

They would always pick on someone at random and say things like, "Did you iron that shirt with a brick?"

On reflection, this was a mental torture as you had to wait, furtively glancing out of the corner of your eye as each man was appraised for his personal appearance and bedspace compliance. You knew that a single failure could result in different levels of response. For example, some errant bristles on the face having not been shaved properly would probably just result in a 'bollocking' for the individual and being told something like "You are a star, a f*cking disas-star" and to report to the Corporal DI later for an inspection to show that they could shave themselves properly. A contravention in the bedspace would probably result in the bed being tipped over or bed pack being launched out of the window, but a toilet not cleaned properly could well mean another bull night for everyone.

Soon, it would be your turn to be inspected and you were just praying that nothing was wrong as the responsibility for causing another bull night was horrendous. Eyes straight ahead, trying to keep breathing, just waiting for the DI to move on; when he did, you could start breathing normally again and praying that the rest of the guys would also pass muster. Of course, the DIs knew this and sometimes, as they walked away, they would suddenly come back to you and look at something again. They would then move on again with what I am sure was a hint of a wicked smile on their lips.

After a few weeks of training, most of us had settled into the routine and it all seemed to get easier. But you had to do your bit, especially on bull nights and by then we had our No. 1 uniform issued to us and even tailored to fit us. With this came a shiny peaked cap and a brand-new pair of shoes that were not to be worn until the pass-out parade. The toe caps, especially, were 'bulled' with 'spit & polish' but I thought it easier and more civilised to use tap water in the polish tin lid on a frequent basis so that they gleamed; and you really could see your face in them.

At some stage, we were given a weekend off for a mini break from the training and those that lived within a couple of hours were given a rail pass to go home. The rest of us were staying, but allowed to go into Lincoln on the Saturday by an RAF bus in civilian clothes. Lincoln was the nearest big town and is a city, due to having a fine cathedral. I don't remember having any issues or remember much about the day or evening other than having a few illicit underage drinks in pubs. Usually, there is a bit of anti-RAF resentment from the local lads as we are seen, quite rightly, as competition for the local ladies; and there had been trouble on occasions in the past. With our virtually shaved heads, we stood out as being RAF. I do remember having the Sunday off for the first time and the joy of not cleaning, marching, running or learning about the RAF for a whole day—luxury indeed!

There was one other standout day to remember for me—the Air Experience Day. I had flown in a small light passenger aircraft several times with the ATC and my father, who had a private pilot's licence. I had also been on a BAC One-Eleven airliner to a summer camp in Malta at RAF Luqa, which is now the international airport, but the RAF was aware that at that time, not many people had ever flown in any sort of aircraft. So, a flight was arranged leaving RAF Waddington near Lincoln for us to fly for an hour around England in a Vickers VC10, which was the RAF's main 151 seat passenger plane at the time, and which remained in service until 2013, giving almost half a century of service.

I can remember being amazed that after only about fifteen minutes after take-off, the pilot informed us that we were approaching Bristol nearly two hundred miles away. The hour-long flight really did fly by, and everyone was buzzing after we landed back at RAF Waddington. By the last week of the six weeks of training at RAF Swinderby, we were all very fit—we could shoot an SLR, do rifle drills in our sleep, and we were sharp in our marching drills on the parade ground. Bull nights had been reduced to just weekly events and the daily inspections had become cursory, or even non-existent. The DIs and PTIs had all seemed to ease up on us. It was as if they felt they had achieved what they wanted in getting our full compliance. We were all ready for the RAF life now, and the focus was on the passing out parade to make sure it was perfect.

Like any organisation, the RAF has rules. The law book for the RAF is known as Queens Regulations (now King's regulations), and these provide the framework of how the RAF works from organisational structure and command system down to administration at unit level. They are applied worldwide across all military activities and locations.

At unit level, Station Standing Orders (SSOs) are the permanent rules pertinent to a particular unit (RAF station) for effective management at that location. A breach of these orders will be an offence, assuming they are aware of the rules. An offender will be 'charged' and subjected to service discipline usually by some sort of restrictions, unofficially known as 'Jankers', which usually meant being confined to camp or having to do an additional drill session or, more likely in the RAF, some menial 'fatigue duties' such as peeling potatoes in the cook house. If an offence was of a serious nature, then the charge could be elevated through the command structure and result in a military court martial.

Finally, to deal with the ever-changing minutia of service life on a base, there were Station Routine Orders (SROs) that were temporary and covered events and activities coming soon. Again, failure to comply could result in disciplinary action, which was known as a 'Charge', as not being aware of orders that had been 'promulgated' was not a defence. Having a charge that was proven on your record was indeed a black mark that could impact advancement through promotion.

At RAF Swinderby, the system was in place for low-level contraventions, but not put on the individual's personal record during the period of their training. If anyone had done anything really serious, they would probably have been ejected from the training and discharged from the RAF. There were several of us put on jankers for all sorts of trivial violations that would result in going to one of the mess kitchens for a delightful evening of washing up pots and pans or preparing vegetables for the next day. I can assure you that three hours of peeling potatoes engenders a desire to not get caught doing anything wrong in the future—but not necessarily to not do anything wrong!

The big passing out day arrived, the culmination of the six weeks' training, and knowing that we were about to join the Royal Air Force for real. Our No. 1 uniforms were pressed yet again and our incredibly shiny No. 1 shoes were about to be worn for the first time. I remember it was a cold, damp and breezy February day and so the parade would take place inside one of the large aircraft hangers, and there would be no fly past. Being inside made it harder as shouted orders echoed off the walls and roof, but by now, we had practiced the whole parade so many times, we knew exactly what to do and when to do it.

The parade took place without incident, watched by the parents of the entire intake. Well, except for my parents, as my father had decided it was too far to drive from Torquay for 'just a couple of hours' ceremony. One other lad had no parents there either, but as he had joined from an orphanage, so that was understandable. Naturally, I was disappointed at my parents' absence, but it was not really a shock. The parade was followed by drinks and food in the Newcomers Club, and the food was actually edible for the first time in six weeks. The cooks and chef had really turned it on for once. Then, it was over. Some speeches by the group captain station commander and the wing commander in charge of the training squadrons, and we were free to go on leave before heading to various RAF stations for our technical trade training.

Having packed my things, I was back on a RAF bus to Newark station, still in my No. 1 uniform, clutching my paper rail warrant, ready to be converted into a rail ticket at the ticket office and then to get a train down to my home in Devon. Although it was over forty-five years ago, my memory of that six weeks at RAF Swinderby is still quite clear. During research, I found things online from people who enjoyed the experience, hated it or felt they were bullied. There was even a reported case of the death a few years after I was there of a recruit pushed too hard during a cross-country run and then a cover up over the circumstances. I have no reason to doubt that it happened and there is also no doubt that we were bullied to a degree, suffered humiliation at times, and pushed way harder than many of us had experienced before. But did it harm me or make me stronger? I was more prepared than most due to having been in the ATC and having gone on several summer camps at different RAF stations, but also because I had an ex-RAF drill instructor for a father who made me his personal recruit to re-enact his national service role over several years.

Thanks to him, I had been bulling my school shoes since I was twelve years old; I made daily bed packs and he inspected my bedroom at the weekend, wearing white gloves. If I did anything wrong or got home late from school, not in time to do my chore list, I received a few lashes with his bamboo cane, which was highly probable. His party trick for visitors was to blow three times on a whistle downstairs in the house and if my two half-brothers and I were not stood to attention in front of him within ten seconds, then we were in trouble. My father was a successful driving instructor in Wolverhampton at the time, with his own business.

He also had two old Rolls Royce Silver Shadow cars that were hired out for weddings. My stepmother, Sylvia, and he would dress up in grey chauffeur uniforms and drive those cars, often getting even more money from tips as people did not realise that the chauffeur owned the cars. It was my job to clean and hoover the cars each Sunday so they were ready to go the following week. How I came to hate confetti, which seemed to get everywhere in the car and was a nightmare if it had been raining and got wet. As a child, I was certainly no angel and I did naughty things; my paper round had a few extra 'private' houses on the end of my round that the newsagent shop was unaware of.

I also got caught shoplifting in a supermarket, and at Codsall Police Station, Sergeant Carter let me off with a warning as he knew my father would, as he eloquently put, "Kick the shit out of you if he found out."

I was so grateful, and it worked, I have never stolen so much as a pencil since! Sergeant Carter was right as I had also stolen some money from my dad's wallet to take a couple of my friends that lived in single parent homes on a day trip on the train to Shrewsbury. I may have got away with it if one of the friends' mothers had not telephoned my father to thank him for his generosity. That got me sent to my room and told to await thirty lashes of his bamboo cane on my buttocks as soon as the Star Trek programme had finished. So, I waited quietly upstairs in dread until I heard the theme tune and prepared for what was coming. The cane session was a bad one, and even though he was angry, he eased off when the welts on my bum grew large and were likely to split. I think I only counted eighteen whacks by then.

Obviously, he was frustrated by this as he said, "Go to bed" and "Goodnight," and then punched me unexpectedly with such force in the solar plexus that it took me ages to get my breathing back to normal.

He was a schoolboy boxing champion so he could pack a punch and I once watched him chin a much taller man on the front drive and break his jaw; he was a work colleague who was leaving my dad's driving school and taking his driving students with him. I think the one thing my dad did approve of was my entrepreneurial streak. As well as working in a garden nursery laying turf at weekends and during school holidays and having my paper round, I also had a small car washing round business, which I had built up doing bob-a-job week in the scouts by asking those that gave me jobs if they would like me to come back and wash their cars.

At 13, my voice had still not broken and so, at Christmas, I would go carol singing like a poor version of Aled Jones. But I would do this on my own door to door, and if I only had to sing a few verses of 'Away in a Manger' or 'Good King Wenceslas', I was fine. One guy asked for a different carol and then laughed at my perplexed face, before paying me anyway 'for my cheek'. My paper round was a mix of working class and middle-class homes and one Christmas Eve, I had received no Christmas tips at all. So, I went back to the newsagents and got a copy of each of the different newspaper titles that I had already delivered, and went to each house, knocked on the door and asked if I had forgotten to deliver their paper, which obviously was not the case. Unsurprisingly, they all said "no," at which point, I wished them a very Merry Christmas and that resulted in all but one customer telling me to wait whilst they got me a Christmas tip. I took the spare papers back to the shop and rode home considerably richer!

When I left the RAF after twenty-two years, I went on to have a successful twenty-year business career and, perhaps, if I had not joined the service, I would probably have been quite wealthy. However, perhaps the RAF shaped me into a person ready to succeed in business by giving me a good work ethic, morals, can-do attitude, and other skills that I may not have had if I had not joined up and gone straight into a business career. What is undeniable for me is that the RAF saved me from a life that could have been determined by a challenging childhood. Also, when I look back, I find my happiest memories are from my time in the service and not my business life, and so I have no regrets about joining and serving for so long.

My father was obviously a parenting outlier to say the least, and if Ester Rantzen's 'Childline' had existed, then I would have probably had the number on speed dial and reached out for help; but, at the time, I just felt unlucky, though I know as an adult what he did to me as a child was cruel and damaging. I could not wait to join the RAF and was so happy when he released me at seventeen years old by signing the paperwork. Luckily, my two half-brothers from his second marriage had a much easier time of it and I'm told that he mellowed a lot in later life, and I have forgiven him as I believe nurturing hate internally causes cancer and illness, but I have not forgotten, and I chose to have nothing to do with him for most of my life.

By 1978, integrated flights of males and females were introduced. RAF Swinderby later gained a bit of fame when it was used in 1987 for filming some scenes of the film Full Metal Jacket; but in 1993, the RAF School of Recruitment Training was moved to RAF Halton and the base was closed, and the land sold off to become a new village and a business park; only two of the large aircraft hangers remain. However, I can imagine on a cold dark winter's night the ghostly crunching sound of marching boots can still be heard faintly in the air. Those few weeks were transformational in my life and many other RAF personnel, as demonstrated by the fact that there is a Facebook Group called 'I was at RAF Swinderby' and it has over 14,000 members and is still growing at twenty new members per month!

2. Learning the Armament Trade

I should probably explain at this point why I decided to become an armourer, or to choose in the correct terminology back then, to become an Aircraft Mechanic (Weapons). Well, the truth is that I didn't really. The RAF Careers Office in Exeter did! I applied to be an Aircraft Airframes or Propulsion Technician—Direct Entry, which would have meant that after three years' training, I would have started my RAF career as a Junior Technician (JT), which was three ranks up from where I eventually joined. But I only had two O-levels and after checking, the RAF Careers Office would not accept my four CSE Grade 2's, which meant only mechanic level entry was available to me. So, for that level, I chose: 1) Airframes, 2) Propulsion (aircraft engines) and 3) Electronics—nothing whatsoever to do with weapons or armament.

By early December 1976, my application was in, but there was a problem. I had put down my stepmother's details on the application form as I had no idea where my natural mother was. They called me at home and said they needed her details for the security check and so I gave them all I had, which was a name and her date of birth. The very next day, they called me back to say they had tracked her down and had her life bio. She was born in Durban, South Africa, in 1936 and because it was a British Colony until 1961, her nationality was not an issue. They then reeled off that she had lived in London, Derby, Wolverhampton, and was currently living at the top of Scotland in Thurso, Caithness, and that no security issues had been flagged up. This was very impressive research in twenty-four hours, considering it was pre-internet days and demonstrates the investigative powers the armed forces had even back then!

So, my bags were packed, and I was ready to go on 4th January, but the day after the Christmas break on 3rd January, came another telephone call. Unfortunately, there had been a mistake and my application for Airframe Mechanic could not proceed as they had no vacancies. Sadly, my other choices of Propulsion and Electronics were also unavailable. However, the good news was I could still join the next day as an Aircraft Mechanic Weapons and if I didn't like it, I could transfer later to one of my original choices. I was so relieved that I could still join up and leave home. To be honest, if they had said the role was a full-time potato peeler, I would have probably gone for it, and so I said yes immediately.

Being under eighteen, I needed parental approval. My father agreed but only on the condition that I signed up for an initial three years instead of six years, to which the RAF Careers Office staff acquiesced. I suspect that the careers office were probably playing me from the start as they had trade quotas to fill. By calling me the day before I joined, they were maximising the chances of me agreeing to change trades. The idea that you can swap later is a joke, no doubt they allow one or two people a year to do just that so they can say it as a truthful fact, but really, there is no chance.

That night, I read the RAF brochures again and found out that armourers have a really varied role and could be arming aircraft with bombs, missiles, loading aircraft guns or working on aircraft flight servicing. Or I could be responsible for the safe storage and preparation of aircraft weapons in Explosive Storage Areas (ESA), which is the posh name for a bomb dump. It also had one line about care and issue of the weapons used by my station, including rifles, pistols and machine guns. So, you know what? Being an Aircraft Weapons Mechanic did look exciting, and of course, I naively believed that I could change trades if I didn't like it—little did I know what was to come!

RAF Halton near Wendover in Buckinghamshire is about forty-five miles north-west of London and dates to the formation of the Royal Air Force when the Air Board purchased the estate for the Royal Air Force, which had been formed on 1st April, combining the Royal Flying Corps and the Royal Naval Air Service. Earlier in 1913, Alfred de Rothschild had invited the army to use his land for summer manoeuvres. The soldiers were joined by No. 3 Squadron Royal Flying Corps (RFC). With the outbreak of World War I, Alfred offered his estate to Lord Kitchener for military training and by 1916 Halton was covered in temporary accommodation (tents and huts) to house 20,000 troops.

In 1917, there was further expansion of technical training in the RFC, and Halton became the main training unit for aircraft mechanics. Permanent workshops were constructed to house the RFC's many trades. The population expanded and by the end of 1917, some 14,000 air-mechanics were trained. At the end of the war, by November 1918, the station was training 6,000 airmen mechanics, 2,000 women and 2,000 boys at a Boys Training Depot, all supported by 1,700 instructors. The end of World War 1, Military Officer Viscount Hugh Trenchard saw his vision of a permanent Royal Air Force endorsed by Winston Churchill, the Secretary of State for Air, December 1919. Trenchard (later known as the Father of the RAF) believed that the only way to recruit high quality mechanics for the Service was to train them internally. His vision was the recruitment of well-educated boys aged fifteen and sixteen who could absorb the technical training. In January 1922, the first entry of five hundred boys arrived at the school, now named No. 1 School of Technical Training. Marshall of the Royal Air Force, Trenchard's ex-apprentices, went on to form 40% of the RAF's ground crew and 60% of its skilled tradesmen. Trenchard is well regarded by all armourers as he has been quoted as saying, "*The armourer, without him there is no need for an Air Force.*"

Fifty-five years later, in March 1977, the latest batch of Aircraft Weapons Mechanics arrived, and on arrival, I reported to the guardroom and made, judging by his apoplectic reaction, a grievous error by calling the RAF Regiment Flight Sergeant on duty 'Flight'. Using this term for a 'Flight Sergeant'. I would learn, would be totally acceptable for the rest of my twenty-two-year RAF career. Maybe it was a 'Welcome to RAF Halton, we are a training camp' thing, or the guy was just having a bad day, or perhaps he was insecure. Anyway, it was not the best way to arrive at a new place.

Over the years, I grew to feel a bit sorry for the RAF Regiment guys as they were not highly regarded within the RAF, who mainly saw them running the guardrooms of bases, or defending perimeters being the RAF's army boys. It was thought that you did not have to be very bright to be a 'Rock ape', as they were known, and use the term themselves with pride. To make matters worse, the army and the marines that I met over the years thought that the RAF Regiment was a joke, so basically everybody looked down on them.

Once in Cyprus, watching the Rocks standing on the top of tables in the bar and raising their heads into the ceiling fans in order to stop them, did make me wonder about their intelligence. I'm not sure that is still the case as the equipment the modern soldier must operate in modern warfare can be quite complicated. The Rock apes also provided the annual General Defence Training (GDT) at every RAF base, which was a refresher of our basic training in terms of shooting our rifles to an acceptable accuracy, delivering first aid, firefighting and nuclear and chemical training drills, which included an unpleasant trip to the gas chamber. The gas chamber used the riot control CS gas or tear gas in pellet form. This is harmless, but will make you cough violently, sting your eyes, make your nose run and, in extreme cases, induce vomiting. You wore a green NBC (Nuclear, Biological and Chemical) or Noddy suit, which was lined with charcoal to reduce radiation penetration, over your combats; and it was baggy enough for you to operate and run about in. Your face was covered by a rubber S6 Respirator and you had cotton liner gloves and rubber over gloves to protect your hands. You also had rubber over boots to protect your feet.

The training in the CS chamber was to officially give you complete confidence in your NBC equipment and your ability to safely decontaminate yourself using chalk pads to dap over your head, face and hands if you were exposed to chemical agents in a war situation, such as an enemy delivering chemical bombs near your location. It was possible, if you did your drills right and held your breath properly when you had your mask off and then blew out hard when you replaced it, to not smell any CS gas at all. So, that did give you confidence that the NBC kit worked. But, of course, that would be no fun for the Rocks. So, they also made us all take off our masks for a minute and as the Rock DI staff tapped you on the shoulder, you had to open your eyes and recite your rank, name and service number before hurriedly leaving the chamber with stinging eyes, runny nose and coughing (thereby inhaling even more CS gas and exacerbating the situation). If you were lucky and it was a windy day, it would blow the CS gas residue off you quite quickly. The temptation to rub your eyes was immense but would be very painful if you did.

Sometimes, in future years after the annual test, despite knowing that my decontamination drills had been thorough, in the shower after a chamber test, I would get a little whiff of CS coming from my hair as I washed it, and if that had been a real chemical agent, then it would probably not have had a happy outcome. I used to tell myself that the whiff had come from the minute standing there like a lemon with my S6 mask off, but it did put a niggling doubt in my mind about the NBC kit. I used to think that making everyone suffer from the thing we were supposed to be avoiding was counter intuitive, but my research shows that the military all over the world seem to see value in CS gassing their service people. Perhaps, knowing that after our trade training, we would all come to resent Rocks; they were getting their retribution in early!

The guys on the course were a mix of ones that I had gone through basic training at Swinderby, such as my mates Tony Wilson and Paul Betteridge, and some that had been back-flighted at either Swinderby or Halton, mainly for medical reasons. The accommodation at Halton was, once again, in large eighteen-man roomed 'H blocks', but with far less hassle such as bull nights, which became monthly affairs. We still marched everywhere though, but this time as a class, and so, we joined the training conveyor belt of turning teenage boys into the tradesmen the RAF needed.

We were soon equipped with general engineering skills such as metalwork, electrical looms and soldering connectors and drilling and tapping holes; and, we had to make test pieces to a set template that demonstrated our abilities in these techniques. This was carried out in the long workshop shed 'lines' that had been built at the time of the formation of the RAF, when Halton first opened. As well as the engineering training, we also started to learn about servicing small arms and the workings of aircraft armament equipment, such as the Avro 7 store (1,000lb bomb) carrier from the Vulcan bomber and the Carrier Bomb Light Stores (CBLS-100) that was used for fighter/bomber jets training bombing sorties, and could be fitted with either four 3kg Practice bombs or two 28lb Practice bombs to simulate dropping 1,000lb bombs as 'slicks' or 'retarded' bomb types.

In addition to preparing bombs with fuses, we also learned about the unguided 68mm SNEB rockets and their Matra 155 launchers. Later, we would be trained down at the airfield part of Halton to learn about marshalling aircraft and flight line activities and requirements. Maybe it is due to memory decline with age, or the fact that basic training at Swinderby was such a life changing event, I don't know, but I do remember the six weeks at Swinderby far more than the four months I spent at RAF Halton on my trade training, in terms of daily life. My abiding memory of each day at Halton is marching down the hill from our barrack block and Airmen's Mess, crossing the main road and marching down to the workshops. Going up and down again at lunchtime before heading back up the hill each evening at the end of the day's training. The biggest difference was that we had evenings and weekends to do what we wanted and there were no curfews or lights-out rules.

So, for the first time, we could have a social life. My eighteenth birthday was about six weeks after I got there and so now, it was even legal to drink in the pubs, not that we hadn't been drinking in them as soon as we arrived. Of course, some pubs were less stringent about drinking laws on age than others and the one that became 'our local' was the furthest away at the far end of the pleasant village of Wendover, the local village near the railway station. It was called the Shoulder of Mutton and had the external red brick façade and a Tudor black beam look and lovely oak beams inside the bar. It is now advertised as a classy food pub as a Grade II listed eighteenth century building, which was not a selling point for me at seventeen years old.

A silly game developed, in which the Shoulder of Mutton had to be renamed each week and if you used its real name or the last made up one, then you had to buy the first round of drinks. Over four months, the establishment acquired numerous names that got ever more ridiculous and the only ones I can remember are: The Dong of the Sheep, the Dog's Bollocks, The Cat's Knackers, and the Crocodile's Cock. Isn't it funny how young men are always drawn to genitalia? One evening, on the way to whatever ridiculous name the pub had been given that week, one of the lads had to call his mum on her birthday. Three of us were waiting outside the telephone box on the camp as he dutifully made his call, but before long, we were getting impatient and signalling for him to hurry it up. His mother was unaware of this and presumably happily giving an update on all the family affairs and we were getting impatient to go and sample the mid-week delights of Wendover. The pub also had a regular two-piece band that sang 60s and early 70s classic pop and country songs that I enjoyed at that time. Our mate finally cracked and told his mum that some guys were waiting for him to go to the "Dog's Bollocks, I mean Cat's Knackers..." and a couple more of the unofficial names of the pub.

Totally flustered at his mother's shock at his language, he just said, "Love you Mum, bye" and hung up.

Naturally, we enforced the rules on misnaming the pub, but I think we let him off at just two rounds of beers. The biggest draw to the Shoulder of Mutton for me was the fact that they would serve me despite me looking about fifteen years old. I have always looked younger than my actual age and am told I still do, now in my sixties, which is great, but looking young at seventeen was a real problem. In fact, when I was twenty-three and married, I was still getting ID checked in some bars!

Looking so young was also a massive barrier to attracting members of the opposite sex as older teenage girls are after young men, not guys that look like their baby brother. I realised that I needed a different approach to 'chatting up' the ladies and found that by being funny and making them laugh gave me a chance for them to get to know me and that seemed to work. I met Linda, my first real girlfriend there. It was a strange platonic relationship as we never had proper sex, which was all my problem, not hers. Every time we got close to the act, my father's words (yes, him again) came into my mind that what I was about to do was wrong and my excited member would suddenly deflate into a limp object and cause me great shame, and, no doubt, great frustration to Linda.

My father once told me when I was about eleven or twelve years old that I should not really exist as I was "the result of the first slapper that let me into her knickers." That kind of statement from someone who is supposed to love you was devastating to me at the time and, with hindsight, damaging. He was not religious but, of course, I saw getting a girl pregnant out of wedlock as the worst possible thing you could do. I was also a bit clumsy as a child and unlucky; things like the main stopcock to the house would break off in my hand when I tried to close it when we had a major water pipe burst. That earned me a smack around the back of the head and being told that I was 'useless' yet again. After a while, that message sticks if you are told it often enough as a child, and it undermines personal confidence and makes you timid and scared of messing it up again.

I have to say that Linda persevered with me, and we enjoyed each other's company. We found other ways of me giving her pleasure, but it never really felt right and, not surprisingly, as soon as I left Halton, the relationship ended. However, I do have a legacy from my last night in Wendover with Linda. We were walking through town and another girl started slagging Linda off for going out with an RAF guy. Some words were exchanged and, stupidly, I got involved and before I knew it, I'm in a fight with the girl's boyfriend, who was far more proficient at it than I was. Clearly, I did not inherit my father's boxing prowess. So, it turned out that my last day at RAF Halton was about dealing with both the RAF and civilian police with my missing front tooth, smashed lip and black eye and slurping soup as it was too painful to chew. Apparently, the guy I had ill-advisedly got involved with was known to the police as a local hooligan, and he went to court and was found guilty and fined £50, which I thought at the time was rather lenient, but then again, I did get involved and his criminal record could well have had adverse impacts on his employment for several years.

On our final day, the training school bosses got us all in a large room to announce that we had passed the training and to tell us our postings; as you can imagine, they were not very pleased to see my beat-up state. Their view was probably 'if you are going to get into a fight with the civvies, at least win the damn thing!' As everyone's name was read out and where they were being sent to serve properly in the Royal Air Force for the first time, the excitement mounted. They called out each name in alphabetical order using our new rank of Leading Aircraftsman or LAC; LAC Adams—11 Squadron, RAF Binbrook. Then, LAC Betteridge—Explosive Storage Area, RAF Coningsby; LAC Davies—Weapon Servicing Flight, RAF Lossiemouth; and so on, until it was my turn.

Now, several weeks earlier, we had been given a class on where we could be posted throughout the UK. I had paid great attention to the instructor and especially the large map pinned on the wall showing all the operational RAF units across the UK. The form allowed three choices for either names, RAF Squadrons, RAF Stations or regional counties or areas; and, under each selection, was a line to give a reason for your choice. There was also one section for you to add a negative choice, so where you definitely did not want to go. So, to me, it had a sort of need to be pragmatic and logical about the choices.

I also figured that if you were not too specific, then you stood a better chance of getting nearer to what you wanted. How naive was I? I put down: 1) RAF Coningsby—reason, I want to work on Phantom jets. 2) RAF Binbrook—reason, I want to work on Lightning jets. 3) Lincolnshire—reason, RAF Coningsby and Binbrook are located there. For my negative choice, I put down Scotland just because it was so far from my home in South Devon. My thinking for the third geographical entry was that it backed up my desire to go to Coningsby or Binbrook and that there were other big stations such as RAF Waddington and RAF Scampton in Lincolnshire with Vulcans and Victors as backup, so I believed I had made it clear that I really wanted to be on a fast jet squadron and in Lincolnshire.

Then, it was my turn to learn my fate. There was an unnerving pause followed by, "Where is LAC James?" I put my hand up.... "Ah yes, the punch bag! Where did you apply for?"

I told him what I had put on the form in my new-tooth-missing, painful lisp voice.

"Well, you have got what you wanted, lad...well, sort of.... LAC James, Weapons Engineering Flight, RAF Swinderby, which, by the way, was your third choice of Lincolnshire."

Now you might have expected sympathetic groans from the other lads in the room that had become my supposed mates and colleagues, but we had already learned that in the forces, you don't do fake sentiments or enable others to feel sorry for themselves. So, instead, there was a roar of laughter and shouts of disbelief—Swinderby!! I was shocked, embarrassed, and wondered if my day could get any worse! The really irritating part was that they were right as I had put down Lincolnshire and Swinderby was in Lincolnshire! It had never entered my mind that I could get posted back there at all, let alone so soon. So, the Personnel Management Centre (PMC) at RAF Innsworth had technically, on paper, another satisfied airman, but I was absolutely devastated and disappointed.

Of course, with time and experience in the forces, you realised that you are just a serial numbered warm body to fill a vacant post in the RAF system. If you have the right skill set and qualifications, great; but if not, then you will just be sent on a training course to obtain them or taught through 'On the Job training' (OJT) at your new location. At your annual assessment each year, the same posting option form is used (or 'Dream Sheet'), as it was known, and I never really took it seriously again; I only once got a posting that I had requested. Years later, I attended a briefing by PMC on how their posting system worked as part of the new enlightened RAF of the 90s.

One of the presenters gave an anecdote about how one person had put down RAF Chivenor in North Devon near Barnstable with the reason as "I have always wanted to be at RAF Chivenor". His second choice was "RAF Chivenor again" with the reason as "You can see how much I want to be at RAF Chivenor". His third choice was RAF St Athan in South Wales across the Severn Estuary with the reason "On a clear and sunny day, I will be able to see RAF Chivenor across the sea."

We all laughed and inevitably someone asked, "Did he get posted there?"

The deadpan reply to which was, "No, RAF Lossiemouth," which is in northern Scotland, about as far away from Chivenor as you can get in terms of operational flying stations.

I like to think he was joking; otherwise, he had just confirmed what we already knew, in that the 'Dream Sheet' was just that!

3. See the World by Returning to Swinderby

I had to quickly come to terms with the fact that the PMC gods had decided I was going back to 'Swinderblitz' (as it was not so affectionately known by some), but this time, not as a trainee but as a permanent staff. I remember feeling really depressed as I arrived back at the guardroom at RAF Swinderby. Not helped by still recovering from the fight and looking a mess, it was not a great first impression, and my new boss asked me if I was a troublemaker! But I soon realised how different things were when you were permanent staff and not a trainee. For a start, I had a single room in a brand-new accommodation block near the Permanent Staff Mess. The food was great and they would even cook things like omelettes and occasionally 'steak to order' for you. But the biggest difference were the DI staff, Rocks and PTIs—the very people we all loathed for making our life hell, it turns out, were mostly nice guys and very friendly and helpful. In the armoury, the NCO was a Sergeant and one Corporal. There was a Junior Technician (JT) Ian Wilson; and four airmen, two Senior Aircraftsman (SAC) John Buck and, I think, John Fowler from Liverpool; and, the lowest of the low, me as the "newby" Leading Aircraftsman (LAC) along with another LAC Liam Crimmins. Apart from Liam, these guys also had RAF experience and history and had served at other places (including Germany).

The senior NCOs kept quite aloof but the others all quickly became my friends and I soon settled in. The Sergeant called Gordon Parkin was a decent guy and, in hindsight, he certainly did not deserve the angst and grief I gave him over the next sixteen months. He understood how disappointed I was not to be at a "proper" RAF station, or on a squadron, and tried to explain that learning the ropes and getting promoted to SAC (which was automatic after a year with good behaviour and technical performance) in an RAF back water was a good thing. I soon got accustomed to the daily rhythm, which was like having a perpetual work out. Just as I had, the recruits learned drill with wooden DP rifles that were old .303 Le Enfield No. 4 rifles that had been decommissioned; however, they still had to be secured in chain racks, in locked and alarmed storerooms. We would carry four rifles to the external door that was opened six inches on a chain and pass them out to the recruits and take them back in two hours later after they had finished training.

The 7.62mm Self-Loading Rifles (SLR) were issued both for dry classroom training and for live firing at the 25m small arms firing range. Real weapons could only be issued and received through locked external doors through a small hatch that was head high in the steel door. On a busy day, we were physically handling a mixture of over three hundred DPs and SLRs as well as issuing boxes of ammunition and dealing with sacks of fired brass cartridges, which had to be individually checked to ensure no live rounds had been thrown away by mistake before being dispatched for recycling by the ammunition manufacturer. Not surprisingly, we all had rather well-formed biceps and toned upper bodies! The rest of the daily duties was carrying out the periodic servicing of the weapons. For the DPs, it was just about checking if the strapping and weapon fittings were secure and the revarnishing of the stock and wooden fittings. For the SLRs, they had to be serviced in accordance with an army schedule. As our range weapons had heavy usage, they would develop snags, which meant they had to be returned to the factory via Donnington for refurbishment. This meant a 'road trip' every six months or so and a day away from the drudge of daily life in the armoury.

I should explain here that when we moved from Wolverhampton to Torquay the year before, it had disrupted my extra year in the school sixth form in order to take some O-levels, and so the plan was to attend South Devon Technical College for six months to finish my studies and take the exams. The teachers at the college were great and, for the first time in my life, I enjoyed learning. However, I needed transport to get there each day and it was decided a moped was the answer.

"Don't worry," said my father, "I will sort that out for you, and you can pay me back when you start earning money."

And sort me out he did. At that time, having a moped meant you rode around on what were essentially 50cc stylish mini sports motorcycles, such as the Yamaha FS1-E (or the 'Fizzie', as they were known). They were speed restricted to 31 mph but if you knew the right garage to 'derestrict' it, then 45mph was possible and even faster if the rider lay flat on the tank and had a following wind.

I had a mate from a wealthy family and he had a Garelli GT50, which looked like a proper motorcycle; he was the envy of the class.

So, when my father did the big reveal on what he had purchased, I was not expecting the Puch MS50D in a green and cream livery. This was not the successful, quite trendy Puch Maxi moped, this was a pragmatic working bike with an engine that was technically a moped used by postmen. It had a huge brown padded seat, a fuel tank that looked like a bloated hot water bottle and spoked wheels with wide bicycle tyres. To complete my humiliation, it had two exhaust pipes the size of drinking straws and proper bicycle pedals and two tiny cream-coloured panniers on the back that would hold nothing useful. I was devastated; how could I be seen riding to the college on something as uncool as a postman's moped? My father was shocked at my disappointment and explained that it was German, so it was well built (it was actually Austrian), and that it was reliable, easy to maintain, had great fuel consumption figures, and could do 120 miles on a gallon of petrol. He had made the selection and purchase totally on a pragmatic criterion, not considering the needs and sensitivities of a sixteen-year-old teenager trying to impress.

Upset as I was, when I rode it for the first time and was moving without having to physically pedal, I loved the immediate sense of freedom it gave me, and whilst I did hate it for being so ugly and old-fashioned, it never let me down, and at least twice, I had to give my Garelli owning mate a lift home when his trendy bike broke down.

Fig 2. An example of a trendy moped (Courtesy internet search)

Fig 3. An example of my not so much trendy Puch moped – Summer 1976
(Courtesy of an internet search)

I was working in the local hotel, washing dishes at weekends and during the holidays and saving everything I could. At last, just after my seventeenth birthday, I was able to trade in the Puch for a glorious Honda 125cc motorcycle, which was my first proper bike. God, how I loved that machine, and that summer I rode it all over Devon and Cornwall with my favourite day trip being a ride up through Dartmoor to Princetown and back, or around the coast road to Lyme Regis.

Fig 4. At last a ‘real’ motorcycle – Late Summer 1976

My father was a keen motorcyclist and owned the now iconic Honda 400/4, which would soon become a classic; he did a great job of teaching me how to ride properly, especially handling and getting good positioning going in and out of corners. He was also teaching me how to drive a car and that was not so good, I was just not a very good driver. One day, two letters arrived simultaneously from DVLC (or DVLA, as it is now called). I was to take my motorcycle test in Newton Abbott at 11:00am on 30th September 1976, and my car test three hours later the same day. As my father was now a driving examiner at the centre, the chief examiner for Devon would do my tests to prevent any conflict of interests.

The big day came, and I rode my Honda with its 'L' plates tied on to the test centre and met the examiner. He started the test and followed me driving around town and then out into the countryside before doing a few test activities. We then headed back to the test centre for the written test part and debrief. Apparently, I was one of the most natural motorcycle riders he had ever tested and, yes, I had passed with flying colours. I rode home in an ecstatic mood and then came back with my dad driving his car as he tried to give me last minute tips. This didn't work and just made me nervous.

Out came the chief examiner again and said, "Right, just do as well as you did this morning, and it will be fine."

But it was an awful test as I made some basic roadcraft errors. At the end, I was waiting to hear the dreaded 'failed' word, but instead he passed me, explaining that he knew that I knew the roadcraft stuff as he had seen it with his own eyes that morning. He told me that if he had not seen me on the bike, then I would have failed the car test. So, thanks to doing my motorcycle test first, I had passed two tests on the same day!

Despite my love of motorcycles, during a trip back home to Devon on leave, I decided that I wanted to buy a car by selling my Honda 125cc motorcycle. My father offered me a loan to give me enough money to supplement and make a car purchase and the loan he offered only had a generous 4% interest, so, 'better than the bank', he told me. The other condition was that, as he knew about cars, he would select the car to be purchased, which was fair enough as I knew bugger all about cars and he had always serviced his own cars.

I should have remembered the moped purchase and realised that something was afoot as he had form in this regard. I was thinking about maybe a Mini, Ford Capri or Cortina or perhaps a Vauxhall Viva. What I was not expecting was a bright orange tangerine three-wheeler wedge shaped microcar oddity, known as a Bond Bug 700ES built by Reliant. Unlike with the moped, rather than being distraught at the complete madness of this vehicle, I was actually intrigued. It looked futuristic, like something out of the TV series Space: 1999. A two-seater with the 701cc engine with 29bhp sat in between the driver and passenger, the whole canopy door lifted, and it incorporated the windscreen and side windows so you could get into the moulded padded seats, and it felt low down as the vehicle was only just over four feet high. Also, with a glass fibre body and low weight of 400kgs, it could power to a top speed of 76mph; it was comparable to a Mini and the Hillman Imp, but I soon found out that going over 60mph could be quite scary and the lean-in corners made me instinctively take them slower than the car could probably have gone. It had a tiny boot and was so impractical in many ways. Due to the engine coming between the two seats, it was utterly hopeless for any romantic activity with the ladies. As they cost £629 new, which was more than a basic Mini 850cc, it is not surprising that they didn't really catch on and production stopped in 1973 after three years with 2,268 Bond Bugs made. Now, they are a collector's item as they are such fun to drive. There are only about 200 left worldwide and can sell for £10,000.

Fig 5. My brothers assessing my father's idea of my perfect car!

I came to really like the quirky little car and it was great for cruising along the seafront of Torquay, but not so good for covering the 270 miles to Swinderby in Lincolnshire, and after a few cooling water pipe issues on the M5 motorway, the four-and-a-half-hour journey took me nearly eight hours to get the Bond Bug to camp. By the time I got there, the front wheel bearings were shot too, so I joined the car club to strip the front wheel hub down and replace the bearing sets. I got used to being stared at in the Bug around Lincolnshire only, but I never dared take it on a long journey ever again.

Being at Swinderby in 1978, I was probably at the physically fittest point of my life. In addition to the constant weapon movements during the day, at night, the other single lads used the fantastic indoor sports facilities to play five-a-side football and use the gym. I had also been given a service issue 'Butchers bicycle' to get around camp, so I raced around on that, too. Despite being a small camp, we had a disproportionate number of PTIs, several of which played football along with young lads from all the other departments. As a result, we had a pretty good station football team and I remember us knocking out RAF Leeming, who were one of the big boys in the RAF Cup; they were not very happy about it.

It was an awful cold and rainy day, and the pitch was soon turned into a muddy quagmire. We took the lead with a breakaway goal early on and then defended like our lives depended on it for eighty minutes. Somehow, we held on for a famous victory. At the end of the game, I could hardly walk and in the bath back in my accommodation block later, both of my legs locked up with a severe cramp. This was turned into a memorable event because the bath water had cooled and I had turned on the hot water to have a soak for longer. When my legs locked up, I couldn't reach the hot water tap to turn it off. So now, as well as my legs twitching in agony, I had very hot water boiling my feet and gradually moving up my legs, threatening to slowly boil my private parts. In the end, I had to use my arms to drag myself over the side of the bath into a heap on the floor!

Naturally, being new to the station, I found myself on Duty Airman duties in the guardroom on Christmas day and I volunteered for an extra day over the Christmas break so that I had New Year's Eve off. But just before Christmas, the fire brigade all over the country went on strike over a 30% pay demand. The strike would last about nine weeks and, in the end, the Fire Brigade Union accepted a 10% pay increase. We were asked if we wanted to volunteer to be a fireman and I jumped at the chance for some excitement. We had two day's training as a fireman at another RAF base on the ancient 'Green Goddess'. It's amazing to think they were used again for another fire fighters strike in 2002!

Then, in January 1978, just a year after I joined up, I was on my way with two other guys from Swinderby to spend a few weeks living in a gymnasium in West Ham in London. I had heard from other RAF guys that the fire strike was boring, and they only attended a couple of incidents or 'shouts' as they were known. Not us, we were in the east end of London, which, in the 1970s, was a mix of industrial areas and suburbs that had long suffered from poverty and deprivation. This was well depicted in Jennifer Worth's memoir, Call the Midwife, which became a book and popular BBC TV series.

Our station seemed to have a crew, out responding to an incident every few hours. I guess, in a way, I was lucky; I was disappointed at the time that I only attended a few house fires, a car fire, and we had to keep visiting a flour plant that had exploded to 'damp down' the storage bins so that they did not overheat and spontaneously combust again. One of our teams were unable to save a little girl in a house fire, and seeing the grown men crying as they returned was a real shock; it took a couple of days for the mood to lift again. We did our best with the equipment we had and basically flooded every place that we visited with water. It was said later that the insurance companies had to pay out £117.5M compared with £52.3M for the same three months the previous year and that they paid more claims for water damage than fire damage. But we didn't care, our mission was to simply save lives and put out the fire. Not being a southerner, I have always had an antipathy towards what I perceive as cocky Londoners, but I have to say the locals were brilliant. People brought us food and gifts, West Ham United gave us free tickets to attend their matches, and the police made sure we found our way to every fire as quickly as the painfully slow and lolloping green goddess fire trucks could take us. Sometimes, striking fire fighters would turn up to watch us work and pass comment on our efforts (not usually encouraging ones). I guess the better we were, then the longer the strike would go on for, and having no income would go on. I remember we turned up in a housing estate to a house fire and my job was always to locate the fire hydrant, lift the access cover and help connect the hose. I found it under a strategically parked car half on the road and pavement. It appeared that the fireman owner was standing across the road with a smirk on his face. I ran and told the policeman who had escorted us what was happening, and he immediately told four of us to roll the car over.

We looked at him and he smiled as he said, "Roll it on to its roof and teach the bastard a lesson."

As we lifted the car, the striking fireman came running over, saying he would move it, but it was too late; the car was on its side. The hydrant cover in the road was now accessible and we paused, but then we heard 'keep going' from the policeman and we all heaved again to put the car on its roof on the road.

The fireman was livid as the policeman told him, "Sir, if that is your car, you had better get it moved as it is now causing an obstruction on the highway."

More expletives came from the fireman as he surveyed his upside-down car with a crumpled roof and side. Apparently, this had happened on a couple of other 'shouts' but after that night, we never experienced it again. To be fair, many of the firefighters were striking reluctantly and when there was a fire at the St Andrew's Hospital in Bow, the real firefighters left their picket lines to help. A few weeks later, the dispute ended. I don't know who paid for it, but a big celebration night was put on in the local community centre hall. There was free booze and food and after, speeches by the good and the great thanking us for our service. I remember that our officer in charge made a speech in response and said that without the police helping us every day, we could not have done the job. The response was all of us standing up and cheering the police present and clapping them. I doubt they have received such adulation very often, but we all knew how vital their contribution was. It must have been quite difficult for them as they work so closely with the other emergency services and, in a way, they were helping reduce the impact of the strike.

Later, we were all sent individual signed certificates thanking us for saving the lives of grateful Londoners and I still have that certificate, all these years later. So, how do you entertain one hundred and fifty young servicemen? Well, of course, you get the strippers in! The young ladies performed their act and then paused awkwardly, holding empty glass Coca Cola bottles. There was a nod from the senior police officer present and the 'act' went to a whole new level of sexual performance. I did not realise that coke bottles could be made to disappear like that!

I think that the process of joining the RAF and being involved in the firefighter's strike gave me a great deal of self-confidence and belief in myself. Back at Swinderby, I soon had a girlfriend and all my performance issues in the bedroom disappeared. *Thank you Karen!*

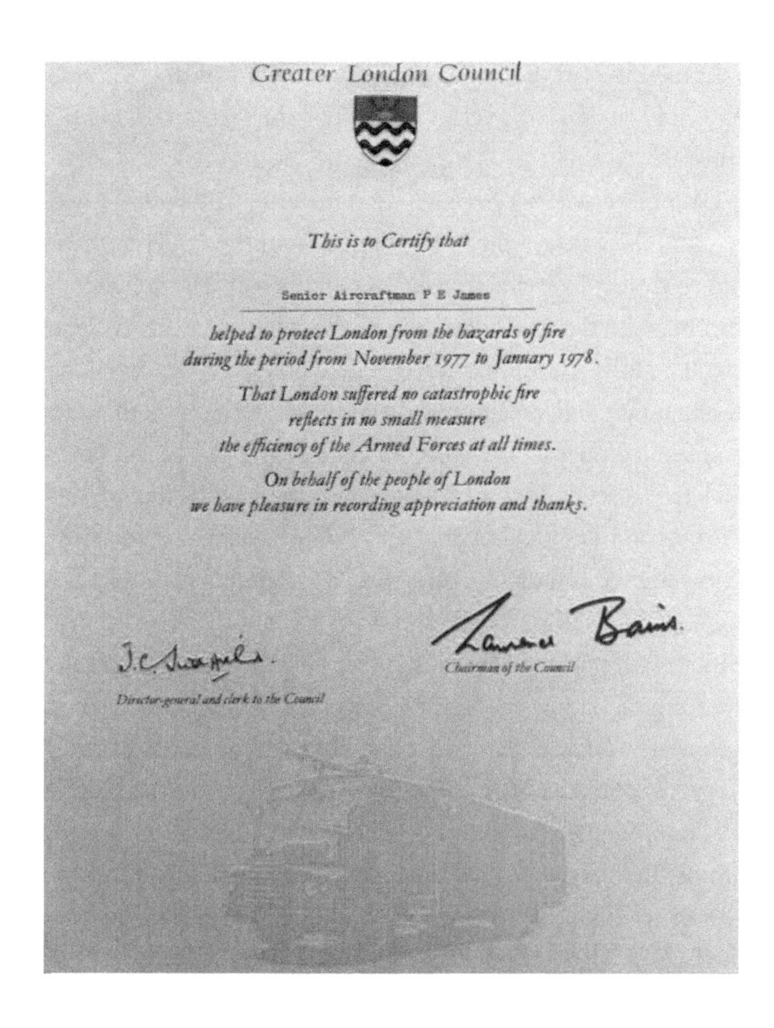
Greater London Council

This is to Certify that

Senior Aircraftman P E James

helped to protect London from the hazards of fire
during the period from November 1977 to January 1978.

That London suffered no catastrophic fire
reflects in no small measure
the efficiency of the Armed Forces at all times.

On behalf of the people of London
we have pleasure in recording appreciation and thanks.

Director-general and clerk to the Council

Chairman of the Council

Fig 6. My Greater London Council firefighting appreciation certificate

I was also given the chance to apply to become NCO aircrew as an Air Load Master (ALM) and so, I applied for that. An ALM is the person that organises the cargo load in transportation aircraft, such as the Hercules, but can also be the person swinging on the end of a hoist cable under a helicopter rescuing downed pilots in the sea or injured walkers in the mountains. I really fancied that job despite it being known as being a 'rope dope'.

I should also mention here something that has bothered me for many years, which is that I owe Sergeant Gordon Parkin a huge apology; he was my boss in the armoury at Swinderby and I gave him a really hard time, even though he was actually very good to me. He had a private pilot's licence, like my father did, and he even took me flying at RAF Cranwell in a Tiger Moth biplane; it was only years later that I really appreciated how lucky I was to have flown in one. He also allowed me time off to spend a couple of days with the Air & Sea Rescue Helicopter Flight at RAF Beverley in Yorkshire to help prepare me for the selection process.

When I went, it was early February and freezing cold with snow everywhere and I was acting as a downed pilot, dropped from the hovering helicopter into the freezing cold North Sea wearing an immersion suit. The helicopter flew away before returning with the ALM dangling below it to rescue me. As he approached and entered the chilly North Sea waters, he discovered that his immersion suit had a leak in a rather delicate area, and he told me all about it in muted screams. The water splashing on my face was cold enough so I can't imagine what he felt! Once I had been 'rescued', we flew low across the East Riding of Yorkshire and chased a fox across a snow-covered field before returning to base. It was an incredible experience and totally convinced me that this was the job for me.

Getting someone selected from the ranks to an aircrew position would have been a feather in the cap for a small unit like RAF Swinderby, and the Squadron Leader OC Administration had kindly given his time to help me prepare for the selection process; we had developed a good but professional relationship. Sadly, despite everyone's help and my enthusiasm for the role, the staff at Aircrew and Officer Selection Centre (AOSC) at RAF Biggen Hill did not agree, and despite getting through the very stressful three-day selection process where people who failed left each evening, I was deemed unsuitable. This would happen to me again in the future, but this time I did not find out until a few weeks later back at the armoury at Swinderby.

I was busy servicing yet another SLR and heartily singing Jilted John's hit at the time called 'Jilted John' but known popularly by the chorus "Gordon is a Moron." I have always had a weakness in that, I find it hard to refuse the challenge of a dare. The other lads said that I wouldn't dare sing the song loud enough that the boss called Gordon would hear it. So, naturally being numb, I sang the chorus "Gordon is a moron" as loud as I could in the echoing service bay, knowing that the boss could hear it very well, but not really accuse me of doing anything wrong. I would just point out that I was still only nineteen years old and obviously very stupid at this time. Gordon's timing was perfect; as my tuneless rendition ended, he called me into the office for what I thought would be a bollocking about disrespectful singing, but instead, he just handed me the official letter about my ALM application failure.

I was devastated and he stuck the knife in by saying something like, "They will never select someone as childish and immature as you."

To me, that sounded like, "You will always be a loser and a failure."

All my newfound confidence evaporated, and it seemed his words were being spoken by my father. So, proving his point perfectly, I lost the plot completely and started to climb over his desk to try and hit him. He was startled and pulled back just as I realised what madness I was doing and stopped myself.

He sat there and said, "Mmmm, threatening a senior NCO? You are finished, James."

And he was dead right, as I should have been. I left the armoury in a right state, "What the hell was I thinking of trying to assault my boss?"

I was panicking and not sure what to do, so I went straight to see the Squadron Leader Officer Commanding Administration that had helped me so much with my ALM application. I knocked on his door, walked in and saluted and then told him all about my stupidity in threatening Gordon, my SNCO, and how sorry I was. I was struggling to hold back the tears as I thought my RAF days would soon be over after an unpleasant stint in a military prison. However, I could not have been more fortunate as the OC Admin officer took pity on me; first, he called the Sergeant, Gordon, who was understandably incandescent that I had bothered this senior officer; and then, I heard him shouting that he wanted me court martialled or at least charged, which was reasonable. Then, the officer made another phone call to someone, before putting the phone down and asking me if I wanted to go to Germany? I was dumbstruck and just nodded yes.

The next day, I had packed my things, arranged for my mate to sell the Bond Bug for me, and then I was on a train to RAF Brize Norton to catch a flight to RAF Gutersloh in Germany. The Sergeant, who must have been bloody furious that I was getting away without any punishment, had been sent on leave, so I never saw him again. However, now, just for the record, Gordon, in the unlikely event that you are reading this, I offer you my belated and sincerest apology for being such a twat to you back at RAF Swinderby. On reflection, being posted back to Swinderby had not been a bad thing as I had grown up a bit and had some great experiences, but I was so relieved to be leaving and going to Gutersloh and not the forces prison in Colchester!

4. The German Eye-Opener

Most of my school age years, during the early 1970s, was spent living in a typical 1960s detached house in a pleasant area of Codsall, which is a large village about five miles north-west of Wolverhampton. A couple of the Wolverhampton Wanderers footballers lived nearby, but this was before the days of them getting stratospheric salaries. Despite living in what most people would consider to be a nice area and being comfortable, we were not immune from living by candlelight during the power cuts, the resultant three-day working week, and general unrest that took place at that time.

The country felt like it was in steady decline, and everything seemed to be getting a bit worse each day. When we moved to Torquay in South Devon in December 1975, and into a house with a great sea view, everything seemed to be getting better, but the UK political turmoil continued and after yet another General Election—Labour's James Callaghan became Prime Minister. Britain also had to suffer the ignominy of being forced to accept a loan from the International Monetary Fund (IMF). Meanwhile, in Cupertino, California, a new company called Apple Computer Company was formed that would go on to transform the way many of us live.

As I travelled by taxi through the German countryside from the airport to the base, I was fascinated by the fact that the taxi was a nearly new Mercedes saloon car, which was far posher than most normal cars in the UK, where minicabs tended to be smelly, dirty cheap cars that were being run into the ground. On my first weekend, I went into the local town of Gutersloh, and I was in awe of the place. For a start, everywhere was spotless; the houses on the way in were tidy and even though winter was drawing in, they seemed colourful with plants and shrubs in their immaculate gardens. The cars on the road and in the driveways were modern, mainly German models as there seemed to be Mercedes and VW cars everywhere. I also noticed on the ring road that had many crossroads, the traffic lights had been synchronised so that if you kept to the 60kph speed limit between junctions, they would be green as you arrived. In the town itself, again the streets and pedestrian shopping areas were spotless, and I was amazed to see glass display cases containing the wares of the shops in the middle of the precincts. I looked around to see who was guarding them. I was even more amazed that evening when I realised that they did not empty them at night! This small provincial German town felt like it was another world to me, so much more modern and brighter than the UK that I had just left, and I fell in love with the country immediately.

RAF Gutersloh was built after 1935 as a German Luftwaffe base and saw active service in World War II as a Night fighter base using the Junkers JU88 aircraft. Legend has it that the tower of the Officer's Mess was used by Hermann Goring, the Reichmarschall of the Luftwaffe, to tell his 'war stories' as a fighter pilot in World War I. In April 1945, the base was captured by the American forces and it was handed over to the RAF in June 1945 as part of the formation of the British Occupation Zone in Germany. From 1968, as Gutersloh was the nearest air base to the East/West German border during the Cold War, it initially had a Quick Response Alert (QRA) role using English Electric Lightning's F2/F2A jet aircraft that could scramble within minutes and climb at an astonishing rate of knots to intercept Soviet aircraft trying to penetrate western airspace. By 1976, the threat had changed to a possible land invasion by Soviet tanks from East Germany and, in response, two Harrier GR3 squadrons (No. 3 (F) and No. 4 Squadrons) had a tank busting role designed to slow down any tank invasion and hold ground until the main US response force arrived. To support this, the RAF Regiment 18 Squadron and two helicopter squadrons were also based at RAF Gutersloh until 1993.

I was working in the Explosive Storage Area (ESA) and, to be honest, it was not very exciting work. We would prepare aircraft weapons such as bombs (live and small practice bombs), SNEB rockets, Sidewinder air to air missiles as well as the 30mm ammunition used in the Harriers guns. But, much of the time was spent uncaging, unsealing and removing the large rubber bag that protected each BL-755 cluster bomb that contained 147 high explosive shaped charge submunitions that were designed to penetrate tank armour. We would carry checks that took about 10 minutes and then reassemble the bag, inflate the seal and refit the cage, and with a wire, lock it to the base and then move onto the next one; there were hundreds stored in the ESA. We were also constantly doing stock checks of cartridges that fired to push the bombs off the aircraft pylons using pistons and every type of ammunition used at the base.

For some reason, often, the totals never seemed to add up properly; so, rather than recount everything again, any surplus items were 'stored' permanently in the deep pond that served as a water storage facility for firefighting purposes. I heard that when the ESA was being handed over to the army in 1993, the pond was drained, and a vast number of explosive items were found in the mud at the bottom! We were also always sweeping buildings and roads and forever painting things like ammunition boxes green.

The ESA was run by the late Warrant Officer (WO) Al Silsby, who had a fearsome reputation and ran the place with a rod of iron. I'm sure he was really a lovely family man at home, but at work, everyone feared his wrath. I had been summoned to his office and it was with great trepidation that I entered his office. Apparently, I had failed to attend the 'Station Commanders New Arrivals Briefing' and the station commander now wanted me charged for failing to comply with SROs. WO Silsby informed me in his inimitable way of shouting so that the entire office building could hear what was going on. But there was something in the way he said that I should read SROs properly as I left his office.

I went to the noticeboard, took down the sheaf of orders with their large bulldog clip, and went to the crew room to read them. Sure enough, there was my name, along with twenty others, being informed to attend the briefing on Friday, 25th November 1978 at 09:00hrs at Station Headquarters. I realised that the Friday was in fact the 24th November. I decided to take a chance to avoid a charge that would have damaged my career prospects by writing a memo to the station commander explaining that on the 25th November, I had indeed turned up at the Station HQ at around 08:45hrs, there was nobody there except the cleaners who told me as it was a Saturday. nobody was working and she thought the briefings normally happened once a month on a Friday.

Obviously, this was risky if the cleaners didn't work on Saturday mornings, but I needed to imply that there was a witness to my arrival. The chief technician in the office said that I couldn't have such audacity to send such a memo to the station commander.

But the WO Silsby walked in, read it and said, "Yes, he can, if that is what happened." As he left the office, he looked at me, nodded, and said, "Mmmm, you will do."

I didn't get charged or even invited to the next new arrivals briefing because I was away in the November, but I did read SROs religiously for the next twenty years of my RAF career. The following week, I saw on the crew room noticeboard 'Volunteers for detachments required'. Now. the wise mantra in the forces is to never volunteer for anything, but naturally, as I was always on the lookout for excitement, I never did follow that advice.

"Where is Belize?" I asked.

"Central America," was the reply from one of the guys.

I was so excited at the thought of travelling to the other side of the world that I went straight to the office and put my name forward. A few weeks later, in November, there were three of us from the winter of RAF Gutersloh on a Hercules C130 transport plane flying to Gander in Newfoundland enroute to Belize. In war films, you see people chatting away in the back of a Hercules and even having mission briefings—that is a fine example of artistic licence. In reality, the noise made by the four propelled engines is so deafening that you have foam ear plugs as well as ear defenders over your ears.

You are sat in a mesh netting seat with a metal frame that, after a while, digs into your legs. The design is to allow paratroopers to sit packed in tightly with all their gear and weapons, ready to jump out of the aircraft via the rear ramp or side door. However, for a six-hour flight across 'the pond' (as the Atlantic Ocean is known), it would be hard to design a more uncomfortable seat. As soon as the Hercules takes off and the 'all clear' sign is given, there is a scramble by everyone to get a place to lie down on whatever cargo is down the centre of the plane, as even laying on a hard wooden box is preferable to sitting in the net seats.

The next issue to deal with is temperature. Depending on where the heating ducts were, you could be boiling hot or freezing cold. With experience, and some resourceful thinking, using whatever clothing you had with you, it was possible to make a bed of some sort. At some stage, a box of usually barely edible 'packed lunch' would be handed out and trading would often take place in terms of the contents, especially the sweets. One thing to be avoided was to take a dump on a Herc flight, as the lavatory was just a chemical toilet with a shower curtain around it; nobody would thank you for doing that, especially those unfortunate enough to be sat close to the toilet area.

When we landed in Gander and disembarked, we met a white winter wonderland. And it was freezing cold. As we were just transiting through, we were not given any winter clothing, just taken to the mess for some food in a bus and then it was back on to the Herc bird for the next leg of the flight. The plane started its engines, but then the pilot informed us that there was a problem with one of them. It was explained to us that the Hercules could fly quite safely on three engines, but it obviously did reduce the safety parameters. So, he wanted us to vote on staying at freezing cold, minus-twenty degrees Gander or taking a bit of chance and risking the flight to Bermuda where the current temperature was a balmy twenty-four degrees and sunny, so a forty-four-degree swing!

Needless to say, the vote was to bravely try and limp to Bermuda and eight hours later, we were being checked in to a posh four-star Hilton in Hamilton and told that all expenses would be covered, which was a costly mistake for the British taxpayer!

This time, we had absolutely the right clothing for the environment, and I found myself sat by the pool, drinking cocktails and eating gorgeous food that I did not even recognise. I was nineteen years old and felt like I was on the TV series Dallas, enjoying the high roller lifestyle. The next morning, the pilot informed us that the replacement parts for the engine were being dispatched from RAF Lyneham but that we would not be travelling for another three days. So, we were given a lump of cash each to spend and that, whilst the hotel was still effectively all-inclusive, we could no longer order room service, had to eat in the main restaurant, and the only alcoholic drinks allowed for free were beers and soft drinks. Apparently, the bill for the previous twelve hours for fifteen guys was in the thousands of dollars!

It eventually took four days to get the Herc serviceable, but then we were on our way after a very pleasant and unexpected holiday interlude. That all soon faded from our minds when we arrived at Belize Airport, bounced our way in the back of a three-tonner lorry along the dreadful roads to the British army camp, and were given our metal-framed bed, complete with mosquito net and locker in a domed metal nissen hut, sleeping with twelve men. The room was a sweatbox, extremely humid; and, the pathetic ceiling fans just moved the hot air around the room and made sure that if someone farted at the far end, you got to sample it at the other end of the room. It was fumigated daily with a stinking chemical smoke and I'm sure that the mosquitoes loved the stuff as they were still there every night, buzzing around, trying to find a gap in our mossie nets over our beds. This was to be home for the next six weeks. If my memory serves me right, I was there with three guys from Gutersloh ESA—Pip Deanelly, Ronnie Mann, and

Billy Barlow. On the first night, Pip complained that he could not get to sleep as he was used to cuddling his wife in bed with his right hand cupping her left breast!

The purpose of the detachment was for us to provide armament support to the detachment of two GR3 Harriers and Wessex Helicopters that had been sent to Belize as a temporary show of force against Guatemala, who have long claimed parts of Belize territory as it had formerly claimed part of the Spanish Colony. As a deterrent, the Harriers flew daily sorties along the border and occasionally fire guns and rockets into the jungle as a warning not to encroach on Belizean territory. The aircrew and ground crew on the Harriers, and we support staff, were rotated on a six-weekly basis between RAF Gutersloh and RAF Wittering to maintain the presence. The base in Belize was being run by the British army and our detachment changeover also coincided with the army swapping over regiments with the Staffordshire Rangers being replaced by the Black Watch from Scotland.

Apparently, the night before we arrived, the Staffordshire Rangers had started their 'farewell to Belize' BBQ by burning clothing hung out to dry by some of the Black Watch guys. The Black Watch guys were obviously not happy about the destruction of their clothing and responded by entering the Staffordshire Rangers' sleeping quarters in the middle of the night and smashing their legs with pickaxe handles as they lay sleeping in bed, which resulted in a couple of broken legs. Needless to say, having noted the severity of the punishment response, we never said 'boo' to those Scottish army guys during our stay in Belize.

I found myself working most days in a wriggly tin shed that was euphemistically called 'The Armoury'; we serviced small arms, SNEB rocket pods, aircraft pylons that carried 1,000lb bombs and other ancillary equipment as well as looking after the Explosive Storage Area, which was a bunch of tents in a fenced off clearing in the jungle. It was hot and extremely humid most of the time but interesting. I was doing a stock check in the ESA, which involved getting on my knees to read the details on the 'kidney plates' on the back of the big bombs. In one tent, I encountered a very large spider, but what really shook me up was being on my knees and looking up straight into the eyes of a huge lizard on top of the stack of bombs; I fell over backwards in shock and I may have squealed a bit. I can't remember.

We would also go from the armoury to the local café that I think was called 'Bettys' (perhaps named after the esteemed café in Harrogate) for a tea break; and, to get there, we had to walk past the fire station, which was another wriggly tin hut. The fireman would often fill up their engine from the static water tank pond that was nearby with a large fire hose. One afternoon, as we approached, the firemen appeared to be extremely animated and were hacking away at the fire hose with axes. We laughed at such ridiculous behaviour until we got closer and realised that the hose was actually a huge snake and apparently a very dangerous one, given the enthusiasm that the firemen used to deal with it!

Every few nights, the cinema, which, as you have probably guessed, was another nissen hut, used to show the latest movies that had been flown out from the UK and for operational briefings from the officers. On the way to one such briefing, I had called in at the shop to get an ice cream and was eating it as I made my way to the cinema. Trying to deal with my rapidly melting ice cream cornet, I did not notice the army officer walking towards me until it was too late. You may not be aware, but the army do not really like the RAF; they don't think we are military enough and call us things like 'Civies in uniform'. This is fine with us, as we think they are lowlife grunts and cannon fodder and it's funny how they always take over our bases whenever we move out of them. However, I had given this army guy a great excuse to bawl at an RAF guy that was desecrating the army base by simply being there and breathing.

Now, when he asked me, "Don't you salute army officers in the RAF?" I did salute him after moving my ice cream to my left hand, but the answer I gave, albeit true, was probably not wise as I used an old joke and explained that we didn't have any army officers in the RAF. There was a pause as he digested this response, which was just long enough for me to realise he wouldn't appreciate it. So, later, I was in front of the base commander, who gave me a lecture about not being on holiday, I was saluting the Queen's uniform, not the person wearing it, showing respect for army officers, et al. Luckily, they couldn't punish me in any way as I had not technically broken any rules, but it was yet another example of young stupid Phil.

Belize City in 1978 was not the trendy and hip holiday resort it presents itself as today, it was a poverty stricken, dirty and rough town. The 1980 American war film 'The Dogs of War' directed by John Irvin was filmed in Belize City to represent an African war-torn town. The seawater waterway that splits the city is called the Collet Canal and Haulover Creek, but the army called it the

Sweetwater canal, and it was so polluted that in the arrivals briefing, we were warned if we fell in and did not get the right inoculations within two hours, it could be fatal. We were also warned which local rums were unfit to consume and that we were not to go into certain bars and whorehouses. Naturally, the list of such places became the very first places we visited, and they really were rough, but we did avoid the dodgy rum brands. We were in one such den of inequity when a drunk guy burst through the door, shouting out a woman's name and brandishing a pistol, waving it all over the place. Lots of screams followed and instinctively, we got down under the table out of the possible line of fire.

He obviously identified the woman he was after as a shot then rang out, she screamed, turned around and proclaimed in disbelief, "You just shot me in the ass, you bastard."

At that point, he turned and ran out the door, the fresh bout of screaming subsided and I found that I had a fit of the giggles; the whole thing was surreal but her response to being shot just cracked me up. We did find a nice bar and restaurant up on the harbour quay side and that became the place we would gravitate to when we bored of whorehouses. One thing about the ladies of the night, though, was that they loved me as I looked so young; they started calling me 'cherry boy' and offering themselves for free so that they had my virginity. Luckily, I managed to resist; otherwise, I would have been in the queue outside the medical hut on base each Monday morning to get a VD test, which apparently was akin to having a cocktail umbrella inserted up your penis and then dragged out to get a sample!

The British army ran a diving training centre at Cay Caulker and we were invited to sign up and attend for four weekends to do British Sub Aqua Club (BSAC), which seemed like a great idea, especially as you left on Friday afternoons, so it was an even longer weekend. We travelled by assault boat from Belize City for about an hour and arrived at the training centre. We were allocated a bunk bed and sent to the classroom to start theory training, followed by a swimming test in the sea.

That night, we travelled by boat to a nearby island that had a hotel and bar, and a good night of drinking with the expected army vs RAF banter. Having travelled back to our accommodation, we settled down for our first night sleeping at the centre. The next morning, at 'stupid o'clock', the army instructors had a unique way of waking you up and shaking off the probable hangover before breakfast. An air horn was sounded at 6:00am and everyone had to leap out of bed, race down the jetty and dive into the sea. The last ones into the sea had to do fifty press-ups on the jetty. Naturally, the army lads didn't bother to tell the four RAF guys about this jolly whizz and how they laughed as we did our press-ups! After breakfast, we had more diving theory lessons and then we were in the sea doing drills such as clearing our masks and doing buoyancy checks. Later that afternoon, we did our first diving in the sea training area of the training centre. I remember having the same sensation swimming under water and not having to come up for air, as I did when I first rode my Puch moped and moved along without pedalling.

After a long day and our evening meal, it was back to the same bar for drinks. However, this time, I took it easy with the booze and the next morning, my body clock awoke me just before the air horn reveille; I got out of bed, went down quietly to the jetty, ready to make the sprint to dive into the sea. The hooter sounded and I was off, sprinting along the jetty, determined to be first into the water.

I was about three quarters of the way down the jetty and shaping up to dive in when there was a loud shout behind me, "Don't go in the water, there is a shark!"

Now that did get my attention, but by then, I was running at full pelt. Desperately trying to slow down, I got to the end and had my toes over the end of the jetty, swinging my arms around like a manic windmill to try and get back on the jetty. I must have looked like one of those cartoon figures after they run off a cliff and try to get back before plummeting down! As I stood there, teetering on the edge and fighting momentum and gravity, I looked down and thought I saw a large shadow pass underneath. The army diving staff explained that a large hammerhead shark had been seen in the diver training area; it had come in close to land as it was probably injured and unable to hunt, so may be hungry. As a result, it was unsafe to dive until it had been dealt with. To deal with the shark, the plan was to get a large ray as bait and then shoot the shark; obviously, the idea of killing two magnificent fish in this way would quite rightly be unacceptable today, but a lot less was known about sharks in those days and their cause was certainly not helped by the 1975 Jaws film and the sequel that was still showing in cinemas at this time.

Several teams went out in small rubber rib boats powered by a large outboard motor out over the nearby Belize Barrier Reef and, very soon, a suitable ray had been targeted swimming along near the surface. All the boats converged on it as did several dolphins who wanted to play with us. Our boat came alongside the ray, which was huge; its wingspan must have been at least eight feet wide. The instructor fired his spear gun and the arrow struck home and…nothing happened; the ray just kept on swimming as if nothing had happened. Next, the instructor bashed it with a paddle several times with no effect. Another boat took our place and when they shot the ray with a second spear gun, this time it was mortally wounded and soon died. I remember feeling sad that something so beautiful had been killed, but I guess we had to deal with Jaws back at the training centre.

After we got back to the training centre, we did some more theory lessons; that was the first diving weekend over and we headed back to the mainland. The following weekend, when we arrived, there was the large hammerhead shark was hoisted up on a gantry. Apparently, the army guys had cut off the ray's wings and used them one at a time as bait. After two days, the shark came in for the second wing and they shot it, so at least it had a quick death. For the next three weekends, I went diving and had completed several dives on the reef; I felt like I was in a Jacques Cousteau programme by the famous French oceanographer as the vast number of different fish, turtles and reef sharks and coloured corals on the reef was just amazing.

After the four intensive weekends of diving training, I had also passed enough training and exams to get to basic diver level under the Br itish Sub-Aqua Club (BSAC) scheme, something that in civvy street could take eighteen months! For our final weekend, we flew to San Pedro. The flight was very short from Belize, but the crosswind made it quite a scary landing. Being able to see the landing strip out of the plane's side window ten seconds before the wheels touch the runway, makes for a bit of a stomach heaving and butt clenching experience. There was not much to do other than sunbathe, snorkel in the sea and drink—what a tough life! I still have a very vivid image of the view from the bar we were sat in, looking over a sandy beer garden; a white wooden picket fence, over the sandy road to the beach with palm trees; and then, the turquoise-coloured sea, blue sky and puffy white clouds. As I drank my rum and coke and the three of us got drunk, talking about all sorts of things, I could not believe where I was, aged nineteen and being paid to be there. One thing I learned though, after getting a painful punch on the nose from Billy Barlow during that drinking session, is that it is probably not wise to get drunk and express your political leanings towards the conservative party to a Liverpudlian and a Glaswegian!

A week later, I was back in Germany at RAF Gutersloh, and it was one of the coldest winters they had had for many years. There was snow everywhere and it was minus thirty degrees. After returning from Belize, I was keen to keep going diving, and so I joined the RAF Gutersloh Sub Aqua Club. Being winter, the weekly meetings were mainly theory classes, some of which I had already completed in Belize. But one week, there was great excitement as permission had been obtained to do a night/ice dive on the Mohne Dam of Dambusters fame. In the planning meeting, it was explained that there was likely to be a layer of relatively thin ice on the surface near the dam itself and being a night-time dive, it would tick off two of the 'experience categories' in our BSAC dive logs. In fact, if, as expected, it was 'nil visibility', then that would tick off a third category. There was even talk of insulated clothing and use of dry bag suits rather than wetsuits as freezing temperatures were anticipated. I was horrified.

All I could visualise was arriving in the dark in a minibus, freezing my ass off carting our tanks and scuba gear down icy steps or grass banks. Breaking the ice and diving down with a torch in pitch black conditions to about thirty metre depth, staring at nothing for twenty minutes, before coming back up and then freezing to death on the surface. I just could not understand the enthusiasm in the meeting to do this ridiculous thing.

At the end of the briefing, I just thought to myself 'F*ck that!' and when I declined to do the dive, I was told that I had been spoilt by diving in one of the world's best diving locations and that this was more typical of British waters diving. It took me a nano second to decide that diving was no longer for me, and I left them to it. I did learn to dive again thirty years later in the sunny United Arab Emirates (UAE) and I even did a week on a live aboard diving boat in the Red Sea and I have dived in Asia and the Caribbean, which was fantastic. But always sunny and with warm seawaters and I have never had the urge to dive in the freezing cold British waters!

The role of the Harriers in a war situation was to fly out of RAF Gutersloh led by 'army intelligence' (which is jokingly regarded in the RAF as a great example of an oxymoron) to intercept the invading Soviet tanks and attack them. As the base was only nine minutes flying time for enemy aircraft and even less for missiles, it was assumed that RAF Gutersloh would be heavily attacked and unusable for the returning aircraft. Therefore, in a war, after the first aircraft launch, we would all jump into our lorries and trailers and 'deploy' to predesignated supermarkets with petrol stations, motorway service areas and anywhere else that a Harrier could land; vertically, if required with a short take-off runway. We would keep doing this for as long as the aircraft returned. When they were 'lost', we turned into cannon fodder foot soldiers to join the RAF Regiment and army guys to try and slow down the advance of the Soviet armoured divisions. By then the US military should have arrived to save the day, so the same war plan as in WW2 then!

In peacetime, these 'deployments' would be simulated by having ten-day exercises in the woods in about a hundred-kilometre radius of RAF Gutersloh, after the army Royal Engineers had spent days putting down sheet metal Perforated Steel Planking (or PSP, as it was known). PSP is an American World War II invention of steel sheets with hooks down one side and slots down the other so that sheets can be joined securely together to cover any amount of ground. The holes not only reduced the weight but also allowed lightning strikes to pass through. Then, the aircraft 'hides' would be covered with camouflage netting as would our accommodation tents, technical support cabins and vehicle parking areas. The Harrier aircraft would then pitch up, hover, then land near the hide, and the war games would begin.

The planning and logistics for these deployments was enormous, especially having to deal with local authorities, farmers and the traffic police; so, they were planned to take place twice a year and were in the calendar with no leave to be approved. Of course, the soviets may be awkward and decide to invade on a different week, so we also had monthly 'no notice' callouts to test our readiness. These would usually start at 5:00am with a siren and tannoy messages on camp and telephone calls to everyone off-base. By 7:00am everyone had to be at their war post, wearing their Disruptive Pattern Material (DPM) camouflage uniforms and have their NBC gear with them, together with field deployment kit, clothing and sleeping bag at the ready. We would then normally get a slightly still warm Egg Banjo for breakfast, which was basically a runny fried egg inside two slices of, if you were lucky, buttered bread. The name 'Banjo' is said to come from the fact that when you eat one, then inevitably the egg yolk breaks, streams out of the bread slices, and onto your chest; and, when you try to wipe off the yoke mess, it looks as if you are playing a banjo! The exercise would often simulate a few air raids to make us get our NBC kit on and then we would prepare all the equipment ready to deploy to war.

At that stage, normally late morning, it would be the end of the exercise or 'EndEx' and we would put everything away and go home. But not this time…it had been noticed that not everyone was bringing in all their gear on a callout and so the station CO decided to teach everyone that did that a lesson and instead of being EndEx and that was it, we were given a deployment destination map and told to set off, which was a nasty shock to everyone. I still remember sitting in the front of a Bedford four-tonner lorry as the escort, going through the German countryside; and, as we drove through a small village, there was an old man stood stationary, almost to attention, on the side of the road. He was just staring as our convoy passed. Our eyes met for a brief second and I had a sudden chill down my spine as the look in his eyes were of pure hatred, of a magnitude I had never experienced before or since. In a second, we had passed and he was gone, but I could not stop thinking about him. World War II had ended thirty-four years earlier, but apparently not for him. I wondered if he was an ex-SS officer and still felt the shame of defeat or something, but I will never forget that look of burning malevolence in his eyes.

It was mid-November and minus thirty degrees at night. An hour later, we arrived at some woods on the army Sennelager ranges near Paderborn and started digging defensive positions and putting up our tents. I was fortunate that all my field gear was stored in the bomb dump and so I had everything I needed; I was still shivering, especially on my guard stint that night. If you took your boots off and left them out in the unheated tent, they would freeze hard, so you either kept them on or put them inside your sleeping bag with you. We heard that some people in other teams got mild hypothermia and that people without their kit were really suffering and, of course, they dare not report it and get into trouble. The next day, the exercise ended, and we went back to base; I bet nobody ever forgot their field kit on an exercise callout again!

Unfortunately, there was another siren sound that went off three or four times in 1979, which signified a RAF aircraft crash. Usually, the pilot managed to 'bang out' to safety using their Martin Baker ejection seat, but sadly, not always. When the siren went off the RAF, police immediately restricted traffic leaving the base, and they would search all the barrack blocks to try and nab single airman to immediately go on crash guard for twenty-four hours until a roster and replacements from the wider-base population were arranged. This was an awful duty as you were either standing guard in some field, taping off the crash site, or, after the initial investigation, picking up pieces of debris and, in the worst-case, body parts.

I was on crash guard twice at Gutersloh and the second time, I was instructed to drive a Land Rover back to camp with the severed hand of a pilot after it had been found, when the rest of his body had already been returned to camp. It was dark and I could not help imagining what the last terrifying seconds of the poor pilot's life must have been like and trying to shake the image of his severed hand in the back of the Rover tapping me on the shoulder.

Whilst the day job in the ESA was often boring, the lads I worked with were always up for a laugh and with people like John Willoughby constantly reciting word-perfectly entire Monty Python sketches; the working day was livened up by us all having fun at work and at play. There were constant pranks taking place and traps prepared in the ESA trying to catch people out and occasionally there would be beaver races. Eager Beavers were military rough terrain forklift trucks to move munitions about and despite weighing about three tonnes, with the driver perched on the top, it could do about 35mph on a tarmac road and about 20mph on the grass areas of the ESA.

Fig 7. An Eager Beaver forklift – Bomb dump racing machine
(Courtesy Francis Tamblyn)

There was a mini-race track setup, but just racing the beavers was not enough; a large tarpaulin was hooked over the towing hook and two-man teams chosen. The race involved two Beavers going head-to-head around a circuit with the co-driver hanging onto the tarp and being dragged along behind the Beaver. Then, we swapped over and did it again, the quickest being the first team to complete two circuits. During one of these races, I was being towed first; I just can't remember who the driver was. Anyway, he got us around the circuit ok, but we were just behind the other Beaver as we exchanged places. I booted the accelerator and nearly did a wheelie start bumping over the kerb onto the grass, but the fast start meant we were now leading in the race. We were bombing towards the U-turn corner near the pond, and I was just ahead of the other Beaver, I realised that as we were on the inside track, if I held off braking until the last minute and turned late, then the other Beaver would be blocked off and must slow down or be forced to go down the side of the pond and be out of the race. I nearly messed it right up by braking far too late; in desperation to avoid driving straight into the pond, I hit the brakes hard and yanked the steering wheel to the left. The Beaver went around sharply and its two right-hand side wheels were over the edge of the pond bank. Luckily, the Beaver just about made the turn, levelled out, and the wheels returned to the ground after bouncing a couple of times, and I accelerated away feeling mightily relieved and excited at the same time.

What I didn't consider was what was happening behind me. My teammate, clinging on to the tarp, was swung out over the pond before being pulled back to the bank; but, of course, the laws of physics came into play, and he had dropped below the level of the pond bank before smashing into it and dropping back dazed backwards into the deep pond. As I crossed the finish line, I was raising my arm in triumph and realised that everyone was bent over, double laughing. I stopped the beaver and looked back at the pond to see a bedraggled and shaken teammate climbing onto the bank. So, sadly, I was not only disqualified, but it cost me several drinks in the NAAFI 49 Club that night in 'compensation' to my teammate, who never raced with me again!

Our games also resulted in the invention, or probably reinvention, of the 68mm rocket tube bazooka. I don't know who or when, but some innovative armourer discovered that if you took the thick cardboard tube that 68mm SNEB rockets came packed in, and secured a Coca Cola tin (other soft drinks in cans brands were available) with the thick strong black tape we called 'bodge tape', you could create an impressive weapon. A small vent was cut into the can and the projectile was a tennis ball, though bodge tape balls could also be used. Liquid lighter fluid was then squirted into the can and the bazooka had to be shaken up quite violently for about a minute before the tennis ball was inserted down the barrel. Then, a lighter was used over the vent and with a loud 'whump' noise, the tennis ball would travel about fifty to eighty yards, depending on the quality of the loading process and the fuel air mixture.

We had two of these bazookas in the barrack block and, after the bars closed, we would sometimes have two teams at each end of the corridor firing these things at each other in a sort of murder ball game. This was fun but, one time, during a field deployment, we were told by our army intelligence officer to expect a night-time attack by a Danish army outfit and that these guys had a reputation for not playing the game, such as pretend dying, when obviously shot by the defenders firing their blank ammunition. Apparently, to make matters worse, they also liked to stick in the boot into anyone they came across during their attack. In an exercise situation, before paint guns and laser technology became available in training exercises, you had to accept that you had been shot if it was obvious and play dead, a bit like when kids play cowboys and Indians.

Our site boss was a young flight lieutenant, but I don't recall his name, which is a pity because he turned out to be a star. He told us that he was worried that things could get ugly with the Danes and it would be better if they were repelled rather than allowed to just run through our defences. He told us to get 'creative', but we were not to tell him about anything we were up to.

So, we got to work and we made about ten of the 68mm SNEB tube bazookas and lots of the round bodge tape projectiles and deployed them in pairs around the perimeter defences, but the nature of the terrain around the bomb storage area was open ground, creating an excellent 'killing field' for defenders; but, on one side of the ESA was a large and densely wooded area—that was where the attack was likely to come from. So, facing those woods, we deployed eight of the bazookas. We prepared the defences of our deployed ESA, we dug trenches with sandbag ramparts, and, in the woods, we used tripwires attached to a Thunderflash, which are like large military banger fireworks designed to simulate a hand grenade explosion. We also set up traps such as twig-covered bear pits (these were only three feet deep and about two feet wide), but in the dark, it would give you quite a fright falling into one. Obviously, as it was just an exercise, we could not put sharpened wooden stakes in the bottom of the pits to impale the invaders; so, instead, we emptied the Portaloo latrines into them so they each had about an inch of the finest armourer's pee and poo in them, which, after a few days of us all living on military compo rations, was rather pungent to say the least.

Darkness fell and sure enough, soon, the air was filled with the loud noise of the double rotors of Chinook helicopters landing briefly in a field about five hundred yards away in a field next to the wooded area. Silence at first, then after about ten minutes, all hell broke loose as the Danes simply charged towards us, which would have been fatal had our weapons been using life ammunition and so not realistic at all. We all open-fired with our SLRs fitted with blank firing attachments, which, as expected, the advancing Danes simply ignored as they were not playing the game, just like when they attacked on the Harrier field hides. However, what they did not realise was that we were not playing by the rules of exercise either.

They were so unprofessional, racing straight at us, shouting like Vikings and firing their guns all over the place. However, as they got within eighty yards, everything changed. First, the Thunderflash tripwires were set off, resulting in loud bangs and dazzling white flashes of light. This made some of them stumble into trees and trip over things; then, the commands to fire the bazookas came, *First rank—fire!* Four of the bazookas were lit and all you could here was 'whump, whump, whump, whump', as they fired the bodge tape spheres towards the woods and the now confused Danes. We had never fired the bazookas in the dark, and the effect was magnificent for as well as a six-inch flame spouting up as the coke can vent was lit. A huge flame came out of the end of the tube as it fired the projectile. What the advancing Danes must have seen was huge flames coming towards them followed by unseen projectiles flying over their heads and landing behind them or crashing into the trees.

Not only was this unexpected, but they also thought we were firing live ammunition at them. Not only did they all dive to the ground, but they then started to run back the way they came or hid behind the trees. Somehow, they had missed stepping into any of our bear pits, but now as they milled about, taking cover, two or three of the Danes got very unlucky and fell into them; there was shouts of surprise and fear, followed by sounds of 'urrghh—hold da kaeft and lort', as they realised what they were standing or kneeling in. As soon as the first row of bazookas had fired, the guys knelt and reloaded with another bodge tape projectile as the helper squirted in a fresh load of lighter fuel. Then, holding an end each, they shook the pipe from left to right to create a gas and air mix.

As soon as they had knelt down out of the way, the next firing order went out, *Second rank—fire!* And four more bazookas lit up the night sky and sent projectiles hurtling towards the panicking Danes. It was like a scene from Zulu with the two ranks firing and then reloading automatically. When the fourth volley of shot was fired, it was all too much for the Danes, and they retreated. I say retreated, but they ran away as a disorderly rabble with another of their guys discovering yet another one of the bear pits on the way out.

We soon heard the Chinook helicopters winding up their engines and the attack was over. We sent out a recce party to check that they had indeed gone and there was no sign of them. We guessed that the Danes would not be happy and, at first light, we swept the woods from two hundred yards out and collected all the bodge tape projectiles that we could find. The soil from the bear pits, which some wit had renamed the 'shitty pits', had been dug the day before and had been used in sandbags to fortify our gun pits—it was taken back to the pits and emptied into them to refill them. There were still signs of disturbed earth under the relayed twigs and grass, but the pits were gone. Sure enough, around 9:00am, a posse of officers turned up and came marching towards us. There was an angry looking Dane, one of the TACEVAL (Tactical Evaluation) Directing Staff, who was a Dutch officer and an RAF Squadron Leader from RAF Gutersloh as their escort.

We stopped them at our line of defences and our flight lieutenant rocked up and said, "Good morning, gentlemen, Can I help you?"

The Dutch TACEVAL officer explained that they were there to carry out an investigation regarding the heliborne assault the night before and reports of the use of live ammunition and health concerns over our defensive trenches.

The boss put on a superb, perplexed look and asked, "Sorry, but what the hell are you on about?"

The Danish officer said something about flame throwers and human excrement in the trenches.

Our boss replied, "No, sorry, I still don't understand. What flame throwers and what trenches do you mean?" And, as the Dane's face got even redder, the boss added, "As far as I know, the RAF, or even the British army, have not used flame throwers since World War II; and, as you can see, the only trenches we have are these gun pits here." He said pointing to the defensive gun pit behind us. "As for human excrement in our trenches, why would the guys do that when we have toilets over there?"

The three visiting officers stepped away and had a private confab and then excused themselves before walking into the trees and looking around. Then, they must have seen the newly laid grass and earth where the trenches had been, but decided against asking us to dig them up to see what lay below.

They came back and the Dutch officer said, "I think that there has been some misunderstanding as clearly World War II flame throwers could not exist here." and he gave the Dane an annoyed look. "As for the human, how you say, excrement? Well, that must have been animal droppings from the boars and wolves in the woods, don't you agree? So, I think we are done here gentlemen?"

The Danish officer managed to compose himself and addressed our boss, "So, what do you suggest that I write in my investigation report?"

And, with an instant reply, and a straight poker face, the boss replied, "Perhaps Danish army Nil, RAF Armourers One?"

That caused sniggers from our guys standing behind me, a smirk on the RAF Squadron Leader's face, and even the Dutch officer had to suppress a smile.

As for the Dane, he was redder than ever and clearly furious, but just said, "Thank you for your cooperation, I look forward to meeting you all again."

And, with that, he turned and stomped off back towards his vehicle with the Dutch TACEVAL officer chasing after him, trying to pacify him.

The RAF Squadron Leader just looked at us, smiled, and said, "The Danes are bloody furious about being humiliated last night," followed by a broad grin and a "Good one chaps, well done! Oh, but no more of the poo traps, please!"

We were later told that another attack on the ESA had been in the TACEVAL plan, but that the Danes refused to attack us as we had refused to give away our secret weapon and that we used dirty tricks. Our officers were delighted, of course. When we got back to camp, the late-night bazooka corridor murder ball games took on a new dimension – 'lights out murder ball'. So now, you were blinded by the massive flash of flame and couldn't even see the tennis ball coming at you at 40mph!

At RAF Gutersloh, the education centre was running voluntary basic German Language lessons two nights a week over six weeks to equip us to deal with the basics for living in Germany beyond 'Zwei bier bitte'. I signed up and this turned out to be far more useful than what I had learned in German lessons at school; it gave me confidence to go and try it out in the shops, but more importantly, on girls in the bars.

At that time, the British Forces Broadcasting Service (BFBS) television service was still getting established and consisted of recorded UK television programmes with soap operas, such as Coronation Street, broadcasted about two days after they had been in the UK and VCR players were an innovation yet to go mainstream. So, people were more likely to go to the cinema than watch TV at home; in fact, not many forces people even owned a TV. But, as a result, there were house and flat parties nearly every week and that, combined with a regular drinking venue routine every weekend, meant I had a hell of a social life.

Due to the tax-free situation, the alcoholic spirits in the NAAFI bar were so cheap that double shots were standard, and you often paid more for the mixer soft drink than the alcohol in the drink and so drinking spirits was more cost effective than drinking beer. After several different drinks, I realised that my favourite was dark rum and coke, and I found that unlike with beer, I had a high tolerance for drinking large amounts of it.

At least once per month, there would be a 'beer call' in the ESA or armoury and that would usually start around 3:00pm on a Friday and go on for a couple of hours; after that, the German civilian contractors bar would open until 6:00pm and the NAAFI bar in the 47 Club opened at 7:00pm. So that natural break gave us time to get cleaned up and get our 'glad rags' on, and if there was no party to go to, then we would descend on the NAAFI bar. When that bar closed at 11:00pm, it would be time to dive into taxis and minibuses to head off into town or to a nightclub called 'The Farmhouse', which is now a trendy jazz club in Harsewinkel, but back then, was a disco nightclub favoured by local German ladies. By now, I had sometimes been drinking for nine hours and I dread to think what those girls had to put up with in terms of me trying to chat them up with my limited and probably incomprehensible German.

It's an odd thing, but I've never been able to pull 'all-nighters' or stay up too late without having a power nap for an hour at midnight so I could keep going; it's as if my 'battery' needs a small recharge 'top-up', if I want to party into the wee small hours. I would explain this to whichever unlucky girl I was with and when I awoke around 1:00am, if they were still there or came back to me on the dancefloor, then it usually meant a good outcome for the evening. I even had a steady German girlfriend for a few weeks, but if I'm honest, I was far more interested in her brilliant Opel Manta car than I was about her. I never wanted to take her on any social events with my mates and when she took me to her home to meet her parents, it was rather awkward as we could hardly communicate with each other. In the end, she asked me why I always just wanted to go out for drives in the countryside; I felt bad then and ended the relationship that day.

On Saturday lunchtimes, we usually went into Gutersloh and met work colleague friends with their families in the restaurant of a large German department store called Hertie, for lunch, before visiting bars in town that played English football commentary on the radio. Then, in the evening, unless a house or flat party was on, it was rinse and repeat with the NAAFI 47 Club disco and Farmhouse night club again. Sunday mornings, if I was on camp, would be a walk down to the Malcolm Club around 11:30am for a late breakfast, tonnes of filter coffee and to buy and read a newly delivered UK newspaper to catch up on the sport.

There were regular NAAFI functions, but strangely, the main weekly event was a disco each Sunday evening run by a guy called Chris Duke, who, as a DJ, had the capacity to fill the floor in one minute with a disco banger only to empty it the next minute by playing some totally crap song. However, despite his dubious DJ skills, the Sunday NAAFI disco was always a busy night with lots of married couples coming in from the off base married quarters for a night out. It was always a laugh and got a bit messy despite everyone being at work the next morning. We have lost Chris now, but his memory as being both the best and worst DJ at the same time lives on.

One of the most memorable weekends for me was over the Easter holiday. A few of us decided to have a weekend in Amsterdam. We boarded the train at 7:30am in Gutersloh and immediately, I heard the hissing sound of beers being opened. I am not a big beer drinker; the stuff does not agree with me, and can make me violently ill the next day if I have too much. So, I just nursed a couple of beers for most of the journey until I could hit the rum and cokes later. When we got to Amsterdam, we checked in at a cheap and cheerful hotel called 'The Albatross' and then we did Amsterdam the way young men all do Amsterdam—basically, no cultural activities, just touring the bars and sex shows, one of which was hilarious when the stud guy could not keep up his erection, which was no doubt made worse by having the audience laughing at him. That evening, we fully investigated the legal brothels in the De Wallen red light district. In order to protect the not so innocent, I will not go into great details here, other than to say we all got very drunk. I am very grateful to the American tourist who found me laying on a canal bank with my shoes and socks off and my feet dangling in the water. The only word I could say was 'Albatross' and somehow, he not only realised it was the name of a hotel, but also produced a map, worked out a route to it and guided me there in person—thank goodness for the kindness of strangers.

The next day, we headed back to Germany on the train. At the border, the German border staff gave us a really hard time as we were drinking beer and didn't have our passports but instead our RAF F1250 ID Cards and NATO Travel Orders, which were perfectly legal but meant that the border staff had nothing to stamp, which obviously irritated them. We eventually got back to camp just in time for Chris Duke's Sunday disco and so the merriment just carried on. As the night ended, someone had the really stupid idea of going late night swimming in the closed station pool as a dare. The swimming pool was due to open for the season on 1 April the next day, as, in the RAF, that is the date winter ends and summer starts regardless of the actual weather and temperature, probably good for planning and budget reasons but still crazy to everyone living in the real world.

So, just before midnight, about six of us went to the pool, which was located adjacent to the Women's Royal Air Force (WRAF) accommodation block and not far from the RAF police station but about six-hundred yards from my accommodation block. We scaled the six-foot-high gate, which had small steel points on top of it, stripped off and jumped into the pool. The pool itself was fifty metres long with four swimming lanes with just a low spring diving board at the deep end. We soon found out that the spring diving board made lots of 'boing, boing' noises when used, and, as a result, very soon there was the sound of police sirens and a RAF plod at the gate telling us to *get out or else*.

Three of the guys left, but the rest of us were in the pool and just hid in the water under the side ledge and kept quiet. Peace returned until one of the guys ignored our pleas to keep quiet and decided he just had to use the diving board again. 'Boing, boing' rang out again and Plod was soon back in his RAF Police Land Rover, but this time he came armed with the pool gate keys and an Alsatian sniffer police dog. I decided to climb out and lay down behind some bushes at the far end of the pool; it was quite cold out of the water, and I was soon shivering. The other two guys climbed out of the pool and stood there naked and were told to get dressed.

Meanwhile, the RAF plod walked around the pool and towards me with his dog sniffing the air excitedly. Luckily, he did not let the dog off his leash, and he came no closer than fifteen feet away. Maybe because I was soaking wet, the dog failed to pick up my scent, and they then walked back to the others. I could not believe I had not been caught as the others were led out of the pool and into the back of the police Land Rover before the pool gate was locked. They drove off and I stood up and then realised that I had been laying in some nettles and had several stings on my legs and lower stomach.

I was on my own, naked and shivering. I got up and went to where I thought I had left my clothes, but they were gone, taken by the police, I assumed. So, I had no choice but to climb over the gate and run back to the block bollock naked. As I landed over the gate, across the road outside the WRAF block, was a young lady canoodling with her boyfriend and she looked on with amazement and amusement as I waved at her and jogged past, I was so cold by now that I doubt I had much on show down below. I got back to my room but could not get in as I had no key, and my roommate was either away on leave or staying over at his girlfriend's that night and so I just had to sit naked in the corridor, just waiting.

Now, I know what you are thinking, and in hindsight, what I should have done was wake up one of my other mates and borrowed some clothes until my pool party buddies returned; but it was nearly 1:00am, so it felt like it was the middle of the night. I was still drunk, though rapidly sobering up and confused about where I had left my clothes. Then, I thought to myself, *Hang on, if the police had my clothes, then even the RAF police would twig that someone else was still in the pool somewhere, so my clothes must still be in the pool, but not where I thought I had left them.*

With this brilliantly flawed deduction, I ran all the way back to pool, naked. As I got to the gate, the young lady and her guy were still hugging but with her looking at me over his shoulder, she waved her hand and wiggled her little finger at me behind his back and smiled wickedly. I smiled meekly and mouthed that it was cold. I climbed over the gate and searched the pool area, looking for my clothes, but they were definitely gone. So it was back over the gate, smile at the girl and run back to the block. Bizarrely, the guy she was with was so engrossed with her that I don't think he ever once saw me. I had just got back in and was waiting for the guys to rock up and hoping that they hadn't been locked up.

Fortunately, Paul Bennellick, known as Banacek, soon rocked up with my clothes containing my room key, laughing that they got off with a verbal warning as he handed me my shirt that he had ripped so he could wear it. He explained that after they had got dressed, my clothes were still in a pile on the floor and the RAF plod had asked who they belonged to. Banacek said that they were his and started putting them on. This was despite him being several sizes larger than me and already having some clothes on! Banacek sometimes went to fancy dress parties covered in green food dye and shredded clothes as the Incredible Hulk. Once at a work BBQ over at the ESA, he was caught out fondling the boss' wife after she had disappeared with him behind the offices.

When they returned together, her white top and white trousers were full of green handprints over her breasts, backside and other intimate places—even the RAF police could have solved that case! But now, my F1250 ID card was missing, and he promised me he hadn't seen it. So, I quickly got dressed and headed back down to the pool to search for it. Fortunately, the courting couple had gone at last and after climbing the gate for the fifth time, I searched the pool area where I now knew my clothes had been for the credit card sized plastic ID card, but to no avail. Losing your RAF ID card is a big security deal, and you must report it to the RAF Police as soon as you realise; it often also resulted in a charge and a £50 fine then, so about £250 now. So, the next day, I duly reported the loss at the RAF police station and the Corporal on the desk called his flight sergeant over.

"So, you have lost your ID card, have you, SAC James? Do you know where you lost it?"

I said I had no idea. He came closer and then produced my ID card and I smiled and said, "Oh, great, thanks!"

However, he then pulled it away and asked how come it had been found on the floor in the back of our police Land Rover.

I racked my brain and then said, "Well, we came back from Amsterdam yesterday and one of your officers kindly gave us a lift to our barrack block."

To which, he said, "Well, isn't that nice? How kind of him," and handed me my ID card. "You are a very lucky boy."

As I left, he said, "Oh, SAC James?" I looked around and he shouted, "In the future, no more midnight swims. So stay away from that f*cking pool!"

I knew he had me and just said, "Yes, flight" and left as quickly as I could, with yet another bullet dodged.

Late one Friday or Saturday evening, I was in a taxi on the way to the Farmhouse, and we came across a girl lying on the side of the road, having crashed her moped. We had been told that RAF personnel should not give medical assistance to German civilians involved in traffic accidents as there were no 'Good Samaritan' rules in Germany at that time, and if, after giving first aid, it went wrong, then under German law, we could be tried for manslaughter. There was another car there and some people just standing, looking at her and not doing anything. I could see that she was bleeding badly from her leg, which was trapped under her moped. I could hear the Rock's warning in my head, but I couldn't just do nothing, and so I decided to get involved and take control of the situation.

In my limited German, I asked the onlookers if anyone had called for an ambulance; nobody had, so the taxi driver used his radio to call his base to order one. I then got one of the other guys to help me lift the moped gently off the girl's leg; then, sobering up fast, my first aid training kicked in with doing the ABC, etc. The onlookers seemed relieved that someone was there and knew what they were doing as they fetched the first aid pillow shaped kit from their car and a coat, whilst I applied direct pressure to the leg. The girl was conscious, but clearly in a lot of pain and going into shock. She was probably also totally confused that she was hearing some bloke wittering on in gentle comforting words in English. Shortly afterwards, the German police rocked up and initially thought I had caused the accident, but luckily the taxi driver soon put him straight on that one. He still took my details from my ID card though and I thought that maybe I would get into trouble. Then, the ambulance turned up and the paramedic guys took over and gave a nod of approval at my efforts.

As the stretcher was pushed into the back of the ambulance, the girl who could not have been older than about sixteen, looked at me, removed her oxygen mask and said, "Danke, sehr, danke."

I knew then that she would be just fine, and I never heard anything more about the incident.

A few years later, the Good Samaritan law was introduced in Germany, and it is now an offence if you could help, but do not do so. You may think it strange that people just stand around in such situations, but it happens everywhere. Years later, on an icy afternoon in January, I came across a car in a ditch in the Cambridge Fens in England with an old lady trapped behind the wheel. The car was sinking very slowly into the mud and water and two other cars had stopped and five people were just standing around. So, again, I stopped and helped, and we managed to get her out before the situation got any worse. At least this time they had mobile phones and they had already called for an ambulance. People fear the situation and doing something wrong and making things worse. I think that first aid training should be made part of the school curriculum for sixteen-year-olds and free refresher training available every three years for everyone. I suspect that with more people trained up and confident about giving first aid, more lives could be saved in the UK each year.

When we were not out socialising, we were in our room trying to keep busy. I was sharing a room with Jez, who became my best friend until many years later, when I got divorced and he became distant to me (these things happen). I could not have had a better roommate as he was and, probably still is, one of the funniest guys I have ever met, despite being from Manchester! Jez was always ready to have a laugh and crack a joke and he was just fun to knock around with. We both loved football; indeed, he played at a relatively highly competitive level into his forties and was a good striker, who reminded me of Gary Lineker in his prime.

I had bought a bright red plastic portable colour TV and we would watch German football when it was on, or just watch German films and their awful music shows—our favourite became a crooner called Heino, who would sing traditional Volksmusik in naff videos set at the top of mountains, in town centres, or bars, and always with his blonde hair, baritone voice and sunglasses permanently on, regardless of if it was sunny or indeed night-time. I later felt bad about that, as I found out that he wore sunglasses due to a medical condition called exophthalmos. He might have sounded truly dreadful to us, but he was a German superstar who sold over fifty million albums, only in Germany and Austria, I presume!

With Jez, like Ying and Yang, came his girlfriend called Jos from Leeds, who he met before I arrived when she was just sixteen years old. Jos had been a cover girl on the station magazine, dressed in a bikini, sat by the station pool that you now know all about. They later married and I think are still together today; and they were perfect for each other and the most 'party-ready' couple I have ever met. Even having two boys did not seem to slow them down much. So, when Jez was not out with Jos, we would be in the room watching TV or reading books.

We both had a passion for Maltesers, that, at that time, came in red cardboard boxes. It is a fact of life that in every box of Maltesers there is at least one disgusting 'soggy' sweet that tastes like you are eating shoebox and we had a game that whilst sharing the box, whoever got the offending sweet had to buy the next box of Maltesers. Naturally, we both soon learned in order to avoid paying you had to try and not gag when you realised you had the offending sweet in your mouth and give the game away, but it was very hard to do! We had an agreement not to play practical jokes on each other, such as making beds collapse when you climbed on it, apple pie sheets, etc. But he could not resist sticking an A4 sized sheet of paper on the ceiling above my bed when I arranged to take a girl home one night and he agreed to stay with Jos.

The poor girl came back, and we snuck into the room; it was her first-time sex-wise, and so naturally tensions were high, and I was trying my best to not let my own excitement get too much and keep her reassured and calm to hopefully enjoy what was about to happen. We were just getting going when she screamed in my ear and pointed to the ceiling where Jez had put up a notice saying, "Big Brother is watching you—so smile." That was all too much for her and she bolted, half-dressed, out of the block, which, after her scream, of course brought half of the corridor roommates into my room to see what was going on. I was hopping around on one leg, trying to get my pants on and remove a condom at the same time, and everyone else is in stiches, laughing at the whole thing thanks, – Jez!

A few weeks later, I was on Duty Armourer duty for the first time, which meant staying the night in the old armoury building. It was a Sunday so I would be in there all day too. I arrived on time at 8:00am and was let in. I knew the guy on duty the previous day and night; he was called Dave and he had been working on his car in the Aden Gun Bay. We opened the double doors and he drove out and then came back in to debrief me on his day of duty and that nothing had happened, and nothing was planned for my day either. I then let him out and locked the door and headed off to the washroom for my morning constitutional. As I sat there in the cubical, I heard Dave's footsteps come into the washroom, followed by a tap getting turned on and the sound of running water splashing into the sink.

I said, "I won't be long, Dave. Will let you out in a moment." As the words left my mouth, I realised that I had already let Dave out, so I called out, "Who's there?" but got no answer, and just silence followed.

I searched the entire armoury building, and nobody was there, and all the doors were secured. I was a bit confused and thought that I must have imagined everything that happened in the washroom. Being the depth of winter, the ancient central heating was going full blast and radiator grills were rattling all over the place and pipes were banging from time to time. Later in the day, I was sat in the office, reading my book, when some movement caught my eye in the Aden Gun Bay. On one of the workbenches were the large cleaning rods, about four-foot-long for the 30mm Aden gun barrels, some had wire ends, others were like giant cotton buds for cleaning after oiling, and I watched as one of these rods that weighed about the same as a two-pound bag of sugar rolled as it slowly moved from right to left, three feet across the workbench.

I stared at the bench and tried to rationalise what I had just seen and thought the vibration from the heating system through the floor and maybe a draught caught the cleaning rod had set it off moving. I hadn't entirely convinced myself about this when, suddenly, having been stationary, the rod rolled back the other way back to where it had started. Now I was freaked out, having just watched something apparently defy gravity. I eventually went back to reading my book and tried to find an explanation for what I had seen, but it just didn't make sense.

Nothing strange happened for the rest of the day, and around 11:00pm, I got ready for bed. The banging and rattling heating system created so much noise that people used the old-fashioned radio in the bedroom to play pop music on Radio Hilversum. The old wooden radio had popular UK stations printed on the frequency chart, such as BBC Light Programme and the frequency of Radio Hilversum, which was the only station available that played English music at that time and was clearly marked on the radio frequency chart window with a heavy red indelible marker line. I put the radio on and went to bed; it took me a while to get used to all the noise and music as well as sleeping in a strange room and bed but, eventually, I managed to get off.

I was startled awake at around 1:00am by a high-pitched electronic screaming sound coming from the radio. I turned on the light and looked at the radio and saw that the frequency needle that had been aligned with the marker pen line was now an inch to the right. I turned it back and straight away, Queen's 'Bohemian Rhapsody' was playing on Radio Hilversum. I went back to bed and turned the lights off and eventually got back off to sleep.

At around 3:30am, the screaming noise was back, but this time, when I turned the light on, I saw immediately that the frequency needle had moved an inch again, but this time to the left of the pen mark. So, no more sleeping for me that night! I sat in the main office with a coffee and started reading the duty log, which was like a diary for the past year, and it soon became apparent that strange things happened in this place. I noticed that the telephone directory sheet under the desktop, which was a Perspex sheet, had at the bottom the padre's home telephone and work numbers, which I had never seen before. I had never seen a ghost or anything remotely paranormal in my life and I certainly didn't believe in the supernatural, but I was being challenged on that belief sat in that armoury office in the middle of the night. I put all the gun bay lights on, drank coffee and read until it got light at 7:00am, when the first of the armoury staff came in to work. He soon told me that all sorts of strange things had been reported in the armoury building and there had even been an exorcist employed in the past to try and clear the place of spirits.

"Well, at least you didn't call out the god botherer padre or run away as some duty armourers have done in the past."

I have never had another experience like that ever, even when I did another duty armourer on a weekday night a few months later. Being a technician and science-focussed, I want to explain things with logic; I do know what I saw and heard but I just can't explain it. A few years later, a brand-new armoury was built and the old one became part of the general engineering flight and was used to service nitrogen bottle trolleys and trailers. The GEF guys also reported strange things going on, so it was not just an armourer thing.

The rhythm of life at Gutersloh became working in the ESA doing mainly boring tasks, just waiting for the social life to kick off at the weekend. Just before Christmas, I met Christina, who had an English father who happened to be a Warrant Officer in Administration, and a German mother and two sisters, and they lived in their own house in Gutersloh. As things developed, we went out together more and more and I no longer went to the Farmhouse nightclub. Our relationship got serious, and I was invited for Christmas dinner with her family, who were all lovely people. A few months later, in the summer, her father said we could take his large Peugeot estate car and caravan on a holiday in southern Germany. He taught me how to operate the equipment in the caravan, change gas bottles and, most importantly, to reverse the caravan into a parking bay, although as any armourer will tell you, if you can reverse two S-Type bomb trolleys with their articulated towing hooks in a straight line and around a corner, then you can reverse anything.

The day we were due to set off, he asked me if we could also take his youngest daughter as she really wanted to go. When I think back, I'm not sure I would be alright with letting a twenty-year-old guy loose with my car, caravan and two daughters, but he was fine with it. We set off for the Koblenz area and it must have been the wettest summer on record as it just rained for most of the week. We did as much as we could in terms of visiting towns, theme parks and walks but the weather did mean we spent hours in the caravan playing board games and keeping the young twelve-year-old sister amused.

Of course, Christina and I had other activities in mind to pass the time and, during one such time in the closed off bedroom, a little voice from the front of the caravan shouted out, "Why is the caravan bouncing up and down?"

As quick as a flash, Christina replied, "It's Phil, he is tickling me."

During the next few days, I 'tickled' Christina whenever we thought her little sister was asleep. We managed to get the car and caravan back home in one piece with only one scare – when the caravan started aquaplaning in a heavy downpour. As we parked up, Christina's mum and dad came out and the little girl ran to them all excitedly to tell them all about the trip in thirty seconds as kids do. However, when she said, "*Phil kept tickling Chrissy lots of times and made the caravan bounce up and down*", Christina and I just blushed and stared at the ground when her dad said, "Did he now?" and looked at me with raised eyebrows and a tiny smirk on his lips, as he enjoyed my intense discomfort and embarrassment. The relationship ended the following year, which was all my fault and a few years later, Christina's dad died of some disease in his fifties and the news really saddened me, as I remembered his trust and generosity.

In September 1979, I was summonsed to see WO Silsby again. With trepidation as usual, I knocked on his office door and was called in. I had been selected for promotion to Junior Technician (JT) and a big pay rise if I passed what we all called the 'Fitters Course' at RAF Cosford, near my old home in Wolverhampton. As I had only done one year of my three-year Germany tour, it could be deferred for another two years, providing I maintained my good annual assessment scores and did not get into any trouble. He told me to go away and think about it.

I was having a great time in Germany and Christina wanted me to stay, but I thought *If I defer and then mess up or get charged, I will lose the promotion.* Also, at that time, as an armourer, if you applied for Germany, you were normally posted there within eighteen months, so I decided to take the offer now rather than defer. It would also mean that after the five-month Fitters Course, I would be a JT only a few months later than if I had joined as a Direct Entrant two years earlier. So, I decided that a bird in the hand was the way to go and took the Fitters Course offer.

Like everyone else, I had taken out a bank loan to buy things we could not afford in the UK, and I had bought a nice Technics Hi-Fi stereo system and a Honda CB360 Motorcycle, and so I had to stay in for the next two months to pay off the loan early. The relationship with Christina suffered as a result and, to make matters worse, I had a drunken one-nighter with some girl I met at the Christmas disco at RAF Cosford. I felt so bad that I told Christina about it when we met up in London. It had been a mistake going to London to try and save a doomed relationship, as it also meant that I could not attend the wedding of Jez and Jos—a decision that I have always regretted. I should have known that *girlfriends are temporary, but mates are usually for life.*

Meanwhile, back at Gutersloh, there was another guy really trying hard to woo her. She eventually married him, so things obviously worked out for the best. I had only been in Germany for thirteen months, but I had grown up to some extent and experienced more things in that short time than some people do over several years, which, for me, just vindicated my decision to join the RAF even more.

5. Homeward Bound

RAF Cosford is located near my hometown of Wolverhampton. It opened in 1938, and as a joint aircraft maintenance, storage and technical training base, it evolved into mainly a training unit, and until 1976, was also home to a large military hospital. One of the dominant features of the base is the enormous Fulton Mess block that was paid for by the local Fulton family and is thought to be the largest accommodation block in Europe. There were also excellent sports facilities at Cosford, including a banked indoor running track that was used for international indoor championships each year. Today, the large hangers are used for the RAF Museum, which has an impressive collection of aircraft and equipment and annually there is a major air show.

In 1979, as well as the Middle East oil crisis, several major events had taken place including Margaret Thatcher becoming the first female UK prime minister; Lord Mountbatten being assassinated by the IRA; and, in the United States, there was a major nuclear power station accident at Three Mile Island. In November, a far less noteworthy event took place as I arrived at RAF Cosford. I was told I was allocated to Training Course AWF 46 and, yet again, put in a barrack block of eighteen-man; but, this time, I was in the Senior Man's room located at the end of it. The main room was full of Weapons Mechanics doing the same trade training as I had done at RAF Halton two years earlier.

I was amazed how young and childish they all were and realised that we must have been just the same and how much I must have grown up in the past two years. I didn't meet my course colleagues until the next day and there was nobody on the course that I knew. I soon settled in; after all, it was just like Mechanic Weapons training at Halton but with additional and more complex topics. We had to march everywhere, of course, but things were much more relaxed, so we didn't always have to march about as a class in ranks of three; we just had to be on time at whatever class we had. I can't remember if we even had a 'Senior Man'. If we did, then it would probably have been John Rourke, a tall

Londoner with a booming voice and an even louder infectious laugh that would bellow out extremely frequently throughout the day.

The training work and play rhythm soon kicked in and we also had more freedom. Weekends were always off-duty and we soon had a Christmas break. It was quite strange to go back to being very close to my hometown, know the area so well and yet not have my family there, other than an uncle and aunt in the nearby Albrighton village.

I went to Codsall, my old village that I had left when I was sixteen; now, of course, I could legally go into the smoky 'Wheel' pub and meet my old friends, which all seemed very strange. It was great to see them but I soon realised how much things had changed in just the three years that I had been away. My best mate at school, Graham Harper, was now a carpet fitter and Jonathan Timmins worked as a panel beater repairing cars in a garage in Wolverhampton. They told me about their jobs, girlfriends, updated me on other friends and it became clear that the highlight of their year was their two-week summer holiday in Spain, and watching Wolves play at Molineux every couple of weeks was what they lived for.

When they asked about the RAF, they assumed that it was about being on parade every day, getting shouted at all the time, taking orders and marching about everywhere. I tried to explain that it was not really like that and told them about Germany and Belize, diving and the adventures I had been on, and that I was now getting promoted again, but it was as if they didn't really believe me; they soon went back to talking about the same things that they probably talked about every time they met. In the toilets, I met another old school friend who had once given me his bike so that I could run away from home. He acknowledged me but said he didn't really like 'stuck up RAF guys' and left me stood there shocked and hurt. In that moment, I realised and just accepted that these guys would never understand me again and that I had left their world far behind me. It was the very last time I sought them out and I have never seen or heard from any of them since.

I got a very different and unexpected viewpoint when I visited the Codsall Laundrette to do my washing as the ones on camp were always busy. I had been in there about half an hour, watching the machine drums go around, reading a book when in walked a girl called Jane, who I had desperately fancied at school. To be honest, I fancied her sister just as much, as they were identical twins! At school, I was certainly not popular but not unpopular either; I was just a little shy, under-confident guy that liked to hang around with the other immature guys that were still more interested in playing football and chasing each other around the school than trying to be cool and chase girls. But when I was about fifteen, the opposite sex was now becoming of interest to me, and whilst I had a couple of girlfriends, like Jill Green and Susan Vaughan, they were really just that—friends who happened to be girls that had no sexual desire towards me whatsoever. It didn't help that I suffered from cold sores and often had several around my mouth and, of course, at sixteen years old, being me, I looked about thirteen.

Thinking back, it was strange that I was really very comfortable having friends of the opposite sex and yet, when it came to having a proper girlfriend, I was hopelessly shy and nervous.

One day at school, in class, I was handed a note from one of the cool kids that said that one of the twins that I really fancied apparently fancied me too but was too timid to ask me out. I immediately wrote a note that was passed to her asking her out. There were howls of laughter as she read it and my note came back telling me that I could go f*ck myself. I had been set up and was naive enough to fall for it. But now, a few years later, things were different; the cold sores had all gone after I joined up and only occasionally came back if I was stressed, run down or out in strong sunlight overseas. I had got taller and bulked out and although Tom Cruise would not get his acting career off the ground for another couple of years, at that time, I did look a bit like Tom Cruise did in *Top Gun* a few years later. Sadly, for me, I do not look like he does now in his most recent reprise of the role in *Maverick*; it is so sad that he has let himself go like that!

As Jane struggled with a pram containing a crying baby into the laundrette, I held the door open, and she did a double take and said, "Philip James?"

Against the backdrop of whirring washing and drying machines, she asked about my life and was really interested in my move to Torquay and life in the RAF and said it all sounded wonderful. I said that she must be happy having a family and a lovely and now, thankfully, sleeping baby, but it turned out one of the cool kids in class had got her pregnant and then wanted nothing to do with her or the baby, other than reluctantly pay her some maintenance each month. She explained how she felt trapped and that her life was over. Soon, her tears were flowing, and I found myself awkwardly holding her and trying to comfort her with words like, *you will meet a decent man and it will all be ok*, which seemed to help. My dryer cycle ended, and I packed the clothes into my sports bag, ready to ride back to camp on my Honda 360T motorcycle.

I said goodbye and wished her luck, and as I opened the door, she said, "You know, when you asked me out at school?" I nodded and internally winced at the embarrassing memory. She continued, "Well, I did actually quite fancy you, so did my sister, but you were just too nerdy."

I laughed and said I totally agreed and didn't blame her at all. Then, she said, "But, I do regret not going out with you now. You are such a nice guy and I wish I was with you."

I didn't know what to say so I just mumbled something like, "Thanks" and left. I don't know what happened to Jane, but I do hope she did meet a decent guy and found happiness for her and her child.

One of the many great things about being posted to Gutersloh was that when living in Germany, in the forces, you got additional money called Local Overseas Allowance (or LOA) to compensate for the additional expense of living in Germany compared with the UK. I think, for me, at that time, it was around £10 per day, which meant an extra £300 salary each month. As most of my money had gone on my social life and with very cheap booze, I had become accustomed to the 'high life', and I just carried on the same way at Cosford, frequenting the pubs in Albrighton and Wolverhampton. Unfortunately, there was no longer any LOA being paid to me as I was now back in the UK and my bank balance soon suffered. My cashpoint card stopped working one day and so I visited the Lloyds Bank branch in Albrighton to try and sort it out. I was invited in to meet with the manager in his imposing wood panelled office and after the usual pleasantries, he asked me if I had my cashpoint card and if could he see it.

When I handed it over, I was shocked that he then said, "We will look after this for you," and promptly cut it up in front of me with a pair of scissors!

I was aghast but he then explained that I was obviously totally out of control financially and I was about £1,200 overdrawn and that things had to change. Now I was embarrassed and felt so stupid. He said that an arrangement would be made in which I could collect so much money in cash each month, and the balance would be kept by the bank to pay off the overdraft over the period of a year. I told him to double what they would keep so I would be out of debt when I left Cosford; he was dubious that I could live on so little, but agreed. So then, 'Party Phil' became 'Stay-in Phil', except for one night a week. And when I left Cosford the following May, I was debt-free and the Lloyds Bank manager praised my attitude and handed me a new card, but with only a £50 overdraft limit; so, he was not trusting me that much! It's funny how you remember people that give you a leg up when you need it or teach you something.

Just like sergeant Carter at Codsall police station, this bank manager had taught me a valuable lesson, and when I saw on my bank statements how much the charges had been for being overdrawn, I was determined not to let it happen again. So, I developed a system I called 'The Book', in which I estimated all my known bills, such as insurance and repairs for my bike, social spend money, holiday money, etc. and averaged them out over a year then added ten percent and divided it by twelve to give me monthly amounts for each item. I then set up 'accumulators' in which the funds would build up and when the bills came along, I had the money ready to pay them. This way, I always had spare money in my account but had to only go by the book and not what my bank statement said I had in the bank.

Forty-four years later, you can do the same thing with pre-paid debit cards by having 'pots of money', but I still handwrite my monthly accounts in a ledger rather than use Excel spreadsheets; and, apart from a brief time during my divorce and after buying a property, I have never been overdrawn again in all that time.

The training course was passing just fine, but I had one thing that I just could not master in the engineering test and that was to file a straight edge on a small block of steel so that when you ran a setsquare along, it there would be no daylight gaps. I felt so stupid and frustrated and I would miss tea breaks, lunch and even go back in the evenings to the workshop to try and master this relatively simple skill that everyone else seemed to find quite easy.

After my filing efforts, the metal block would soon have undesirable slopes on the sides and at the ends. This was going to be a huge problem as there was an engineering test that involved cutting tin plate, shaping holes to fit an electrical connector, soldering wires into it and finally adding a perfectly filed rectangular steel metal block with a tapped hole and threaded stud coming out of it. I could do all of the other tasks but if I could not master this metal block filing, then I would fail this part of the Fitters course. Step forward, the legendary metalwork instructor sergeant Dick Bates, who had tried in vain to help me master this basic skill. Just before the final test started, he called me outside and handed me a block with three sides already filed to perfection, leaving one block side with blue engineers' dye on it. He told me to, under no circumstances, do anything other than file off the blue dye off the one side and then to not overdo it. I just stared at him and nodded gratefully.

He said, "I have never seen anyone try so hard to achieve something and still be so shit at it. You deserve a break, but it's our secret, okay?"

So, I did exactly what he told me, and I got a 95% pass for the project. He told me that the rest of my work with the tin and connectors was so good that he deducted a few marks to make it believable. I still owe that man a beer! I heard that he told future courses of how he had trained this guy Phil James who was bloody useless at filing but never gave up and eventually passed with distinction as a sort of class motivator.

I think enough time has elapsed to tell the truth, and, even now, whenever I am doing any DIY and pick up a file, I think of Dick Bates and say *thank you* quietly to myself. That was the only time I have ever really worked hard at something, but just could not master it; well, apart from mathematics and when I took up playing golf!

A few months later, AWF 46 had completed training and we had all passed and become Junior Technicians (or JTs) and we awaited our posting destiny. You will recall that I had filled the form with the 'dream sheet' and so, it was with some trepidation that I approached the noticeboard to see that I had been posted to Weapons Engineering Flight (WEF) at RAF Marham, which was near Kings Lynn in Norfolk. The next day, I was on my motorcycle, heading east to start a new chapter in my RAF life.

6. The Land of Windmills

RAF Marham was built during the World War I mainly to defend Norfolk from German Zeppelin raids and then as a night-time flying training base. However, as part of the rapid RAF airfield construction programme in the 1930s, it became a bomber station during World War II, and had extra-long runways than the typical station layouts of that time. The US stationed several bomber squadrons at the base until in the 1950s, the RAF based the English Electric Canberra and then the V-bomber force and supporting tankers, such as the converted Vickers Valiant and Handley Page Victors. Eventually, the aircraft of the 1950s would be replaced with the latest Panavia Tornado GR1 strike aircraft, with the last Victor leaving in 1993, after thirty-five years' service, and the last Canberra Squadron was disbanded in 2006, after fifty-two years at Marham. Today, after the base infrastructure was further developed into a Main Operating Base (MOB), Marham is home to the famous No. 617 Squadron of Dambusters fame operating the impressive F-35B Lightning aircraft.

In May 1980, when I arrived with my shiny new Junior Technician four-bladed propeller badges on my arms, the entire airfield was a massive building site as Hardened Aircraft Shelters (HAS) were being constructed, ready for the arrival of the brand-new Tornado squadrons due a couple of years later, which would remain in active service until 2019.

I was employed in the Weapons Engineering Flight (WEF) but not given a role in any of the bays, such as small arms, carriers or the ejection seat bay; instead, I was part of a roving weapons team that dealt with the Canberra aircraft on 100 Squadron and the Victor refuellers on 55 and 57 Squadrons, as none of the squadrons actually had any full-time armourers as there would not have been enough work to keep them busy. This suited me just fine as I would have been so bored just working in a bay all day. Instead, the days were varied, and often urgent tasks came in that we had to deal with. The main jobs were related to fitting and removal of ejection seats from both the Canberra and Victor aircraft, taking on and off the huge underwing fuel tanks of the Victors or removing and fitting the fiddly detonators of the Canberra wing tip fuel tanks.

It was exciting for me to be working on aircraft at last, but they were ancient engineering from the 1950s and it felt like we had old, glass-valve driven radiograms like my grandma had, compared to the newer Harriers, Jaguars and the brand-new Tornadoes that started arriving on the base. Also, despite their bomber heritage, the Canberra was now a reconnaissance plane and target tower and the Victors that once carried nuclear bombs were now air to air refuelling aircraft; neither carried any forms of armament, so they did not need full-time armourers. To me, they were just not proper squadrons. A bit like when I was posted back to Swinderby after training – I felt I was again not in the mainstream RAF and so I also applied to go back to Germany and requested a squadron post. Having said that, the advantage of there being no armourers on the squadrons meant that our team in WEF had to go with them whenever they went abroad on detachments.

There were about thirty armourers on the base and whilst I can't remember everyone, the names that I do recall are WO Jim Gordon, Kev Boel, Dave Bowles, Andy Wilton, Gary Freeman, Mike Hardwick, Dave Kemp, Ken Leek, Ned Nevin, Dave Dixon, Rab Gillespie, Tony Rolfe, Tony Dunk, Paul Botting, Brian Molicon, Al Hewitt, Paul Whitbread, Neil Raynes, Harry Haldane and Andy Povey. Many of the team were single guys that were up for detachment trips to anywhere.

Soon, I had gained sufficient experience and it was my turn to be selected. I would eventually go on detachments to Palermo in Sicily and Goose Bay in Canada with the Victor squadrons and Akrotiri in Cyprus with the 100 Squadron Canberra's. And, boy, did I just love going on these trips!

One of the first things I did after arriving at Marham was to work out my finances with my new 'Book' system, and with my pretty big jump in salary, I worked out that I could afford a bank loan to upgrade my motorcycle. So, off I went to see the bank and, perhaps, surprisingly, they agreed. I had always wanted a Honda 550K motorcycle, which had wicked split flared exhaust pipes on each side, but that was not really a big enough jump in engine size and power to make the upgrade worthwhile.

Within a week, I had a 1,000cc BMW R100/7, which I found in a bike shop in Swaffham. It was not the sexiest looking machine, but I loved the gold colour and black colour scheme, and the way the engine twin pots poked out on each side with those 500cc pistons, which made the whole thing vibrate side to side until you opened the throttle. So, come rain or shine, this was now my mode of transport. However, not for very long.

A few months later, I was at a girlfriend's house party in Kings Lynn. I got drunk after catching her kissing another guy and was leaving in a right huff. She was a casualty ward nurse and because she had seen the broken body consequences of so many motorcycle accidents, she not only hated motorbikes but refused to ever go on mine. It was about 1:00am, and, ignoring her pleas to not ride back to camp, I set off. I was drunk, angry and stupid and rode way too fast and the inevitable happened.

Hurtling along the A47 Kings Lynn to Norwich trunk road, I remember overtaking an articulated lorry just before an elongated sweeping s-bend that went through the village of Middleton. I had ridden this bit of road many times and enjoyed banking over the bike one way then the other, but I had never attempted it in the state I was in and doing over 90mph. I swept around the left hander and the foot pad grazed the tarmac before I levelled it out along the short straight stretch, before banking hard over to the right to take the next bend. I had overdone it, and the footrest and then fatally, the right-hand engine pot, caught the ground and the bike slipped away from under me.

Bizarrely, being so drunk probably saved me from serious injuries and perhaps even saved my life, as I was so relaxed, I didn't tense up for the impact. I ended up on my bum and back, sliding down the road; my gloves had both come off and my jeans and cool grey leather flying jacket took a beating. I was not really concerned about myself as I watched my BMW bike keep going straight along the road but crashing from one side, flipping over and crashing on the other. The taillights were still working until about the third impact on the road, then the bike came to a halt, screeching like some animal in its death throes.

There was silence for a moment then I heard the airbrakes of the articulated lorry I had just overtaken hissing away and the driver trying to stop his vehicle in time. I remember thinking, *great I've trashed my bike and now I'm about to be killed by a truck*. But it didn't happen, he stopped about ten yards behind me and got out to see if I was okay, and telling me he thought I was going too fast into those bends.

Then came a shout from the top window of this standalone house I had ended up outside, "My husband is out on night duty; do you need an ambulance or the police?"

I looked at the house and saw a blue and white sign that simply said "Police." The crash experience and now, the thought of getting done for drunk driving, made me instantly sober and I quickly replied, "No, thank you, but can I leave my motorbike on your drive until the morning please?"

She kindly agreed and the lorry driver and I manhandled the wreckage of my bike into her driveway and kicked the broken bits off the road. He then generously offered to give me a lift to Narborough, which was near RAF Marham and left me at a telephone box when we got there. I called the duty staff at the guardroom, asking for a lift and was helpfully told that everyone in the mechanical transport (MT) section was off duty.

I explained rather forcefully that I had been involved in a traffic accident on my motorbike and was injured with blood dripping from my hands and arms, which was true. Twenty minutes later, the duty medic rocked up in a military ambulance and took me to the medical centre to patch me up. I had enough about me to know not to go anywhere near the guardroom and RAF Police, who may well have breath tested me for alcohol; so, instead, I went back to the block and my bed. Sure enough, at 8:00am the following morning, the RAF plod was knocking on my door, telling me to report to the guardroom to report the accident. I did and just said I lost control of the bike on a patch of oil or something and he was okay with that.

A married mate of mine had a car with a trailer, and we recovered the bike and I also apologised to the lady in the police house for disturbing her. She said her husband was sleeping off a night shift, so I didn't have to talk with him. The motorbike was a right mess, and I got a vacant garage near my block and locked it away as I felt so stupid at what I had done; and the consequences—I could hardly look at the wrecked bike.

Later that day, I girded my loins and went to the garage to assess the damage properly and I realised that, as the main chassis was undamaged, the bike could be saved with a new fuel tank, new front suspension forks, new engine cylinder end pot caps and a new front wheel. I was already paying a fortune in insurance, and I knew that if I claimed it, then my future premiums would go through the roof, so I decided against claiming. I also felt I owed the bike a debt of duty to repair it, and so over the next few months, I sourced replacement parts from as far apart as London and Leeds through bike trade magazines and telephone calls. I remember coming back on the train from Leeds with an entire front fork suspension unit sat between my legs that I had got for £80. I could not get the exact colour match as it had been a limited addition so I used the nearest gold colour match that I could find and sprayed the tank. Eventually, after several weekends working on the bike, it was fixed, and I took it to the motorbike shop in Swaffham for them to check it over and give it an MOT—it passed, and I was mobile on my Beemer again.

Fig 8. My first big bike – after I rebuilt it

I found myself being interviewed by the RAF Police again in February 1981, after having a minor accident in a service vehicle on camp. I had been on duty armourer, which meant that you had to be physically in the armoury until all flying had ceased and there were no small arms weapons going in or out of the small arms bay or until midnight, when you could lock up and go home.

It was a Friday night and so nothing was happening, and I had taken the portable TV from my room in the barrack block on camp into the crew room so I could pass the evening hours watching TV. At midnight, I locked up and took the TV back to my room in the barrack block using the armoury service vehicle. I was then about to drive to the guardroom to hand in the armoury keys before returning the vehicle, which was a Ford estate car, back to the MT yard. Unfortunately, I was tired and not focussing enough and when I reversed, I bumped into a parked Vauxhall Viva that had parked up in the five minutes it had taken me to drop off the TV. The Viva was quite badly dented in the side and the service estate car had a dent in the back and the minor collision had smashed one of the rear lights.

I took the car to MT and the guy on duty said that the damage was too much to 'cover up' and that he would have to report it, which I totally understood. However, on the Monday morning, I was summonsed to see the Warrant Officer in MT and he went at me as if I had burned down the entire MT hanger and destroyed the entire fleet of vehicles. I later heard a rumour that this guy had a real downer on armourers as one of them had apparently defiled his eldest daughter and got her pregnant and then was not willing to marry her. Anyway, he was adamant that I was going to be charged not only for damaging the car, but for misuse of service fuel as I had not driven the direct route from the armoury to MT but had carried out a hundred-yard diversion that was not authorised. He also politely informed me that there was no way I was going on my first detachment (or 'a jolly holiday', as he called it) to Sicily later that week and that I should go press my uniform and bull my shoes to try and make a good impression with the officer charging me. I thought that this was a total overreaction to what damage had happened.

I explained everything to my own Warrant Officer in the armoury and he called the MT guy before putting the phone down and telling me, "Unless you can come up with a good excuse as to why you took the service vehicle to your block at midnight, then I can't help you."

So, off I went and gave it some thought on my way to write the report for the RAF police. I was so deep in thought that I hadn't noticed the three black cars coming towards me down the road until the first one, with the station commander's pennant on the bonnet, had just about passed me and the car behind with the royal emblem pennant on it drew level and stopped. I looked at the car and sure enough, sat in the back was the big boss, the lady who I had pledged to serve, Her Majesty, Queen Elizabeth II. She wasn't looking at me, but looking ahead and probably wondering why they had slowed down and stopped, not knowing the reason was me. I stood ramrod to attention and saluted the side of Her Majesty's car, just as the group captain opened his car door and started climbing out.

He saw my salute and said something like, "I should bloody well think so, you should pay more attention and salute your own station commander too next time!"

He was right, of course, and I mumbled a, "Yes, sir, apologies," as he got back into his car and the cortege drove off.

The royal family often flew in and out of Marham when travelling to and from Sandringham estate. Though I never saw the Queen again, but I always stopped and saluted the station commander's car when flying its pennant if it came anywhere near me!

I was a little shook up by this experience, but by the time I had got to the RAF Police office ten minutes later, I had come up with a plan. I explained to the RAF plod interviewing me, that, having locked up the armoury and set the alarms that I realised I had left my beret in the crew room. Rather than open everything back up again, and despite it being after midnight, I decided to go to my room in the barrack block to get my No.1 peaked cap so that I would not be improperly dressed after returning the vehicle and then walking back to my barrack block, which was about one hundred and fifty yards away. This was a farcical story, but it had a legitimate basis, and so that was what went into the report. The next day, the MT Warrant Officer burst into the armoury to see our WO, Jim Gordon, and there was a heated exchange before he left his office.

He saw me and said, "You f*cking smart, Alec. You think you can do one on me? I will get you for this!"

I just looked surprised and gave him my best innocent 'butter wouldn't melt look' as he stomped out. All charges were dropped and next day, I was on a Hercules C130, heading for Italy.

The detachment was with 55 Squadron Victors at Palermo in Sicily. I don't recall what the detachment was for, but probably to refuel a fighter changeover going to and from Cyprus. I don't remember any significant events, though apparently, I could have lost my fertilisation capabilities or got some brain damage after lying on the apron tarmac underneath a Victor, sunning myself. One of the Avionics ground crew, a trade known as the 'Fairies' came down the aircraft access ladder and was horrified to see where I had been lying. He had been running the Tactical Air Navigation System (or TACAN), which was used before GPS was invented. When running, it would be belting out UHF radiation.

He went pale when he saw me and asked, "How long have you been lying there?"

I told him about ten minutes, and he looked relieved and then explained the potential health issues. I did manage to eventually share in the creation of two boys later in life, so clearly no damage down there, but I know several people who would argue that maybe some brain damage to me had occurred. At the weekend, we managed to watch the Liverpool vs West Ham League Cup final in a bar and, as we had fans of both teams on the detachment, it made for a lively afternoon.

The next day, nursing hangovers, we were all at work to see the Victors off back to the UK and pack up all the equipment, ready to return home. We were not flying until the Tuesday and, with the Victors gone, we were given the day off and so came the highlight of the trip, other than the great Italian food, which was the day trip up the dormant volcano—Mount Etna. It was quite a long drive in the minibus and when we got there, it was too cloudy to see the top of the mountain. So, the authorities had closed it to visitors; we just hung about the visitor's centre and its bar, naturally, for an hour or so before heading back. We now know it was really closed because of increased seismic activity.

We flew back to Marham the next day and that evening, on the BBC Six-O'clock News, one of the main stories was the unexpected and extremely violent eruption of Mount Etna. In the helicopter news reel footage, there was what was left of the visitor's centre, where we had been about thirty hours earlier, poking out of a stream of red-hot molten lava.

Fig 9. Me in Sicily on detachment with 55 Squadron

When we station armourers were deployed with the Victor squadrons, the guys on the squadrons were not unfriendly, but they did not really involve us very much either and they stayed quite aloof.

This was certainly not the case with the Canberra 100 Squadron boys, who, despite calling any of the supporting crew from RAF Marham 'Klingons', did welcome the armourers as if we were their own. I was lucky enough to escape the UK winter with a detachment to Cyprus with 100 Squadron at the end of October and it was quite an experience.

I was trained to help the Flight Line Mechanics (FLMs) with servicing the aircraft, but my time was spent mainly on preparing the banners, which were the large material strip targets towed in the air by the Canberra for the Lightning and Phantom fighters to practice their air gunnery skills. I would be on the team that attached the banner at the end of the runway and recover it from the aircraft after they landed having been shot up.

Fig 10. Me with the 100 Squadron FLMs in Cyprus

We often painted messages on the banners, and I wrote 'Happy' Christmas instead of 'Merry' and got the mickey taken out of me, followed by a Christmas card with the word 'Merry' crossed out and 'Happy' inserted in pen. The message in the card said, "Merry Christmas, you burk, from all the lads on the squadron," which was a nice touch.

Most afternoons, we were often finished by 3:00pm and so we spent the rest of the day down at Buttons Bay or the Diving Club, which were both on RAF Akrotiri base. We went for a traditional meat Mezzeh one evening and I found out just how disgusting cheap Cochanelli red wine is! There was also a very drunken day at the weekend spent in Paphos, and on the way home, I had my backside out of the minibus trying to do a moony, but all I achieved was to lose my wallet, complete with ID card yet again.

Luckily for me, my wallet got handed in to the local police, and when I went to collect it, even my last four Cypriot pounds was still in it. I asked the policeman to give it along with another five pounds to whoever had handed my wallet in as a reward for their honesty. The entire detachment was about working quite hard from early morning until mid-afternoon and then relaxing with alcohol somewhere into the night. I mentioned this earlier about being in the NAAFI Penn Club one evening and watching the Rocks stopping the ceiling fans using their heads and convinced me that their reputation for not being that bright was probably well deserved. A real highlight of the trip for me was having my name selected to fly on a banner target towing sortie in a Canberra. I was super excited as I was strapped into the spare ejection seat, and we took off. I couldn't see anything from the seat and the pilot told me to put the seat handle pins back in and go down into the bomb aimers position, lying down in the nose. From there, I had a great view and whilst I couldn't see the banner getting shot at, the Lightnings came alongside, and I got some great photos.

When it was time to return to base and land, I asked if I could stay where I was and the pilot said, "Okay, but it's on your head if anything goes wrong."

As a result, I have some great pictures of approaching the runway and a surprised look on Finkie's face, who was the FLM marshalling the aircraft in to park on the apron when he saw me waving at him.

Fig 11. My view from the Canberra's bomb aimers position landing at RAF Akrotiri in Cyprus

A few days later, on 7 November 1980, I witnessed the very aircraft that I had flown in—WH667 (or 'Juliet')—crash on take-off. I was on the target banner party as usual and I attached the wire to the rear of the plane, then gave the pilot the 'thumbs-up' sign and returned to the Land Rover on the side of the runway as normal. The Canberra revved up its engines and started down the runway. The Corporal in charge with me was a 100 Squadron engines guy, or 'Sooty' as they were known (I can't remember his name), but he immediately sensed something was wrong.

The plane set off down the runway but did not seem to be going quite as fast as usual and by the time it disappeared over the hump in the runway, the concerned sooty was out of the Land Rover and standing on the bonnet. The Canberra reappeared and was starting to climb but did not have enough power and it banked over. I didn't see the ejection seat go out sideways as it was too far away; then, the plane fell again and disappeared from our view.

Suddenly, there was a huge fireball and a pall of black smoke going up into the air, and what seemed like several seconds later, the sound of the explosion reached us and the sooty was shouting, "No, no, no".

The next thing was bedlam, sirens were wailing and fire engines and ambulances were racing to the scene at the far end of the runway. We drove back to the squadron dispersal; everyone was quiet and in shock. About half an hour later, I was called into the office and asked if I would be willing to go with the Squadron Leader Medical Officer to the crash site as they didn't really want the 100 Squadron guys involved. I agreed, and was soon walking through the debris, some of which was still burning and smouldering. There was a smell of burning oil and, every now and then, a whiff of something I had never experienced before—burning human flesh. I was totally focussed on looking for ejection seat parts that may still have live cartridges fitted and trying to ignore the rest of the debris that we were picking our way through.

We found out later that one of the engines had exploded, which led to a fatal loss of power and the plane banking over. The pilot was a squadron leader 'Paddy' Thompson, and he was supposed to have retired from the RAF a few months earlier but had been asked by the RAF to stay in for a few months longer until his replacement was ready to join the squadron. This detachment to sunny Cyprus was supposed to be his swansong before he left the service, but by volunteering to stay, it had cost him his life.

The evidence in the subsequent crash report suggested that he had not tried to eject, but instead attempted to keep the aircraft in the air so that that his aircrew colleague could eject. The navigator was a young Flying Officer called Mark Wray and he did eject. Unfortunately, as the plane was banked over, he was fired into the ground at ninety degrees and was still sat strapped into his ejection seat, but with his head missing when we got there.

The fireman gave me a set of ejection seat safety pins and when I found the bomb aimers ejection still intact, I made sure that its pins were all still fitted. On the navigator's seat, I checked that everything that could go bang had done so. All ejection seats have a drogue chute, which is a small parachute that stabilises the seat after ejection the right way up. It allows the aircrew to stay with the seat and its oxygen supply until a safe height is reached; then, the occupant can be safely separated from the ejection seat and descend on a normal style parachute with their dingy hanging on a cord below them, in case they landed in the sea. The navigator's seat had its drogue shoot laid out along the ground, which was like a macabre training demonstration layout of how an ejection seat works. However, it took me a while to find the various explosive-containing components from the pilot's seat as these had been dispersed all over the place, along with the pilot's body.

I heard the medical officer say, "Over here" to the station photographer and the male nurse with him, "That is a foot, photo and bag it." Then a bit later, "Here is part of a skull bone—photo and bag that too," and so on.

I had been on a couple of crash sites before with the Harriers in Germany, but this was so soon after the crash and I was actively looking for dangerous components in a crash site that was still burning, and my heart was beating pretty hard by now. Finally, I found the main gun from the unused pilot's seat still attached to the rail, but laying on some still burning grass. I called the fireman over, who put out the flames immediately and then I carefully inserted the safety pin into the main gun to stop it firing.

This process went on for about half an hour and at the end, the medical officer came over to me and said something like, "Well done, lad. This is an awful thing to have to do."

It was decided later that for the 100 Squadron ground crew the 'wake' would take place at the Penn Club, and I went along. Some of the 100 Squadron aircrew came for a while too and some nice tributes were made for the lost aircrew. The navigator was a new guy and so, whilst the lads mourned him, it was nothing to how they felt about 'Paddy' as, apparently, he was loved by everyone on the squadron and the fact that he should have already retired and not been in Cyprus just made things feel even worse.

I would go on to serve on several fighter squadrons in my future career and I would see more aircrew lost in crashes, but I would never see the ground crew mourn a pilot the way these guys did. The solemn 'wake' split, and I ended up at the Diving Club with some of the ground crew when the whisky came out and the 100 Squadron war stories were told long into the night. There was no target banner flying for the next two days and then we were due to fly home anyway. The mood was sombre and quiet for those days, and I was glad it would soon be over as suddenly what had been a great detachment had become an immensely sad affair.

The day before we departed, the squadron aircraft took off, loaded up with boxes of fresh oranges and watermelons that we had bought from the local shops. I think it was a tribute to the lost aircrew and, perhaps, a show of determination not to be cowed, but the remaining four Canberra's did a very low fly-past over the dispersal, banked sharply and then climbed steeply up into the blue Mediterranean sky. They flew past so low that I could clearly see the boxes of oranges stowed in the bomb aimer's nose cone!

After Christmas at home in Torquay, I made a bit of a name for myself during the Goose Bay winter detachment in January 1981, on one of the Victor squadron detachments. I soon discovered that winter in Newfoundland, Canada, was very different to the weather in Cyprus as it was freezing cold at minus 35 degrees, I had never been anywhere as cold in my life. We had thick parka coats, trousers and Muttluk boots that were like huge furry wellington boots and fur-lined mittens. I nearly crapped myself in the taxi going into town on the first night as the drivers just casually slide their cars sideways into bends on the snowy roads.

Soon, the entire detachment ground crew were in the main bar that was noisy and smoky with a soft rock band playing and people dancing. I met a local Canadian girl, a separated woman in her late twenties. Later, I found out she also had a young daughter at home. I ended up going back to her place and stayed the night and the plan was that she would drive me back to the base first thing the next morning. During the night, a huge snowstorm had passed through and when I looked out of the window, there was about three feet of snow everywhere and the roads on this little housing estate were blocked.

The girl, let's call her Mary, as I can't remember her name, said, "Oh right, we have had a bit of snow."

A bit of snow? Having never been skiing, I had never seen so much of the stuff!

"Awe, don't worry honey, they will clear the road in a day or two and we have plenty of supplies in so we won't starve," she confidently predicted.

For me, this was a nightmare. It was a ten-day detachment and I was trapped off base in this woman's house with her young daughter on the first proper day of it. Apparently, the snowstorm was far worse than the met guys had anticipated, but of course, they were used to such things in Newfoundland. The rule was that if you got caught in a snowstorm so severe that you could not see, then it was known as a 'Whiteout', and you had to take cover at the nearest substantial building until it passed. Several of the squadron guys were on their way home from the bar the previous night when the storm struck, and they had taken refuge in the Canadian Air Force Sergeants' Mess.

Mary called her dad, who worked on the base, to tell him that she was ok and could he let the British guys on the Victor detachment know that – "What is your name, again?" she asked me.

"Phil James," I replied, feeling very awkward.

"Yeah, tell them Phil James is safe with me."

It turned out that Mary's dad was the Mess Manager in the sergeant's mess and so he just had to stick his head into the bar to let the guys know.

Apparently, he was not best pleased when one of the lads jeered and said, "That lucky bastard, we are trapped in here and he is banging some bird in town."

I'm sure it got a bit awkward when the Mess Manager pointed out that 'the bird being banged' was his daughter!

Things got worse, as being a military base, they had their roads cleared within hours, the housing estate with Mary's house on it was not cleared until after another less severe storm had passed through. Eventually, after four days of being confined to that house and having to entertain and home-school yet another twelve-year-old girl, I finally made it back onto the base, and you can imagine the reception I got from the squadron lads! Luckily, I did not have to go anywhere near Mary's dad in the sergeant's mess and I decided not to trouble Mary anymore with my company before we flew back to the UK.

In January 1981, Ronald Reagan had become US President; meanwhile, back at RAF Marham, a bunch of the single lads decided to go to the monthly NAAFI disco. For these prestigious events, busloads of girls would be shipped in from Kings Lynn and Downham Market. However, this arrangement sounds far more exciting than it usually was, as not that many girls would attend; and those that did, were not always what you might expect.

However, at earlier discos, we had met a group of nurses on a night out party there and on this Friday, I would meet my future wife and mother of my children. As she has not given me permission to use her real name, I will call her Sam. She was dancing with a friend and so I did the sidling up next to her thing, dancing by myself and trying to get hints from her that she was interested. We danced together for a while and she was very keen to tell me that she was not on any of the buses; she had her own mini waiting in the NAAFI car park. She told me that she was not supposed to be there and that the Welsh guy doing the tickets on door worked with her dad in the Mechanical Transport section, and so she was also a bit nervous as her father, who was ex-RAF conscript and now a civilian refueller driver, had forbade her to ever go onto RAF Marham.

Apparently, he was not that enamoured with the RAF guys he worked with. I still can't think why, but I naturally assume it was because he worked with MT guys and not armourers! We had a few dances and drinks and then she had to go. I said I would meet her after she finished work at an estate agent in Kings Lynn at lunchtime the next day.

True to my word, I got my mate Dave Bowles to take me in and when we went in, she was not sure which one she had danced with the previous night – so much for my Tom Cruise look alike killer looks! The relationship developed and I even bought her a steak meal at a Bernie Inn, so obviously, this was getting serious! When her parents realised this, they were apparently not best pleased, and then demanded to meet me.

At the time, I thought I looked cool smoking the King Edward cigars I had got duty free from Cyprus, but she wasn't keen, so I stopped that. When I found out that I could get the pub landlords to pay me £40–£50 for a box that had only cost me six quid tax-free, I was delighted to give up and pocket the profit. Like most guys in the forces, I had smoked cigarettes after I joined up, but I was in a nightclub near Newark one night and I only had fifty pence in cash left, so it was either a packet of ten Benson & Hedges cigarettes or a pint of Tetley's bitter – the beer won, and I never again smoked after that.

I was soon being collected from the guardroom by Sam in her sporty little black mini car and we were off to meet her family. In hindsight, I should have probably prepared more for this event than just throwing any old gear on and going for a pint on the way to the guardroom. Not that if I had thought about clothing it would have made much difference, as I was really not interested in fashion at all and preferred to spend my money on motorcycle gear; as a result, I had the dress sense of a washing line. To this day, I don't know what I was wearing, but my future mother-in-law never tired of telling the story of how I came to her house with holes in the elbows of my jumper. They say that you only get one chance to make a first impression and clearly, I did that, but not in a very positive way!

Anyway, we went in her living room, which had a roaring open fire and was really hot and so off came my grey leather bomber jacket as the introductions were made. Her dad, called Tosh, was a big bloke, so straight away, I was a little bit intimidated. Then came the mother, Rosie, as she would always be known to me. Rosie is a lovely woman, but when she wants to show annoyance, she can give you a look that would wilt the stoutest heart and that was the look that I was getting right then. Sitting on the settee were Sam's younger sisters, Louise and Jane, and they were looking at me and giggling to each other. So, I sat on the edge of the smaller settee and, after a bit of small talk, was then just ignored and made to feel about as welcome as a fart in a spacesuit. Sam was also feeling uncomfortable and suggested we go to the pub in the village for a drink and I was so relieved to get out of there.

I said my goodbyes, and Tosh, with his brilliant Norfolk accent, said, "So boy, are you gonna come on Sunday and help me with the pigs before Sunday dinner?"

I had no idea about what pigs he was on about and apparently, I had just been invited for Sunday lunch, so I just said, "Yes, of course, I would love to."

As we were leaving, Rosie said, "And try and find a jumper without holes in it," to which, everyone laughed.

Totally confused, outside the house, Sam pointed out the holes in the elbows of my jumper and then showed me in the dark that their garden was actually a smallholding full of sheds that had large sows and their piglets that Tosh fattened up from birth to selling them at the market after about eight weeks. Soon, it was Sunday morning and I rocked up around 10:30am on my BMW bike, which at least scored me some smarty points with the sisters.

I would realise in the future that Tosh liked to 'muck-out' the pigs first thing in the morning, but not this day. This day, he had waited for his unsuspecting hapless helper to arrive. He took me into one of the sheds and there were metal crates with sows laying in them and some had little piglets squealing and climbing over each other to access a teat. He went in and, with his shovel and broom, piled up the pig poo and straw and then I shovelled it into a wheelbarrow, *easy, peasy*. I thought, but then we left that shed and he said that dealing with the small piglets could be tricky with the grumpy sows and so I could clean out a pen with older piglets that had been weaned.

As I opened the wooden door and entered with my wheelbarrow, the stench hit me like a wall and I gagged a couple of times; it was a large pen with about fifteen small pigs in it and they were just walking around in poo slop liquid, which I had to scoop up on a shovel and into the barrow. I suspected straight away that this was a test, and one I was determined to pass. It took me about twenty minutes and several wheelbarrow loads out to the trailer, where it was just dumped by Tosh on top of the load of straw and poo mix to soak in. In the shed, I had streaming eyes and kept gagging if I disturbed a particularly pungent patch. *What the hell were they feeding these animals with!* But whenever I left the shed, I had a smile on my face and even whistled a little tune. I could see Tosh watching me closely and he seemed a bit disappointed.

Back in the shed, I reverted to sniffling and gagging and I had not noticed that he had followed me in, so now he knew how I was really dealing with it, but when we had finished all the sheds, he said, "Urghh, pigs that age, that is the worst you will ever have to deal with, and actually you did alright, boy."

I was relieved that the first dad 'Are you worthy for my daughter' test had been passed. When we had finished, Rosie came out with a washing up bowl full of potato peelings and vegetable waste and told me to follow her. I nodded and thought *now what.*

In another shed, there were some larger pigs. She said, "These will all be off to Lynn market tomorrow, but not this one."

All the pigs had been fed with ground meal from paper sacks and mixed with water but separated from the others was a lone pig that was bigger than the others, and I soon realised why, as he got the washing bowl contents tipped into his compartment floor, and with excited snorts, he set about demolishing them.

"We will be eating him soon," said Rosie. She must have noticed the look of horror on my face as I thought they were going to slaughter the poor thing before lunch, so she said with a laugh, "Not today, you daft bugger! Today, we are eating one from the freezer."

Apparently, in every few litters of piglets born, there will be one with some sort of deformity. At the market, such an animal would not only not sell but could also lower the value of the other pigs with it as buyers would be wary. Rather than dispose humanely of such animals, Rosie and Tosh would give them special treatment, better food and so produce tastier meat. Of course, still not a great outcome for the pig, but at least during the brief life it had been bred for, it was at least treated better than its siblings. This pig had a deformity in his testicles, which were blown into a large single bag.

Later, at Sunday lunch, we had, of course pork, and I had never tasted pork like it; it was so tasty. The vegetables were gorgeous too as they had been grown locally and despite it being the depths of winter and, as Rosie would say, "they were gettin' on a bit." they really were still far superior to anything I had eaten from the supermarket or the mess on camp. The meal went well, and the entire family went for a walk in the nearby woods to help digest the food. I got chatting with Tosh and we seemed to hit it off apart from me disagreeing with him that Norwich City were better than Leeds United!

Over the next few months, Sam and I spent more and more time together and fell in love. Rather than watch Prince Charles and Lady Diana get married, we went out for a picnic somewhere and started talking about the future. All the time, I got closer to the entire family; they were wonderful and normal and cared for each other, so the opposite of my own experience of family life. Tosh took me rough shooting for pheasants with his dogs and lent me one of his shotguns.

There was also a shoot that the farmers ran just for each other and one time you would be 'brushing' or walking through the fields of sugar beet with the dogs to drive the hidden birds into the air and towards the waiting line of shooters. Then, in the next field or wooded area, it would be our turn to stand and shoot at the birds as they flew overhead. This provided a cheap day of sport that others would pay thousands of pounds to do. Of course, those that paid big money expected and got to shoot at hundreds of birds; if these farmers shot thirty birds in a day, after eight drives, it would have been considered as a good day.

The other thing I noticed was that unlike the big shoots where huge numbers of the shot birds are often discarded; here, everyone took a brace of pheasant home with them to eat, so it all felt sustainable. I learned that Tosh was the gamekeeper who would rear about five hundred pheasant chicks each spring and release them in the summer around the farms that our shoots took place. We would also go shooting ducks over at his friend's farm in the Cambridge Fens and whilst I was not that keen on pheasant meat, I did love duck meat, despite the odd piece of lead shot and the immense time and effort required to pluck them properly. The saying, 'you pluck a duck twice' is true, as once you have the main feathers, there is an insulating layer of much smaller feathers to deal with.

Eventually, I approached Tosh about marrying his daughter and he gave me quite a hard time before saying "yes" with a wicked smile. Sam had been engaged for years to another guy before that all fell apart, so she said that she was not interested in having an engagement ring and I was naive enough to believe her. I later learned from Jules, and meeting other women, about just how important the engagement ring, and in particular, the size and quality of the diamond 'rock' is. But Sam did not seem to care at all and so I never gave it another thought. Things were hurried along a bit quicker than Sam's parents wanted, as my application to be posted to Germany had been approved, and in

August, I would be heading to RAF Bruggen to work on 20 Squadron, which at that time flew Jaguars. So, a wedding date was agreed for the following March in 1982.

At work, I was so excited that I was finally going to work on 'real aircraft' and kept talking about it at work, carelessly not realising that in effect, I was putting down everyone that was staying at Marham and working on Victor and Canberra aircraft. In the end, chief technician Gary Freeman called me into his office and explained that he had left the RAF when he was forty years old but hated it outside in civvy street so much that he went to the careers office and asked to re-join until he was aged forty-seven. Normally, this would simply be refused, but his reports mentioned that he had served with distinction and that even on his final day in the service, he had been at work as if it was a normal day. So, they let him back in to serve another seven years. The point he made was that I should never burn bridges or say anything derogatory about where I was working, as one day, I may want to be back or must deal with the same people in the future.

In my head, I was thinking that I never wanted to come back to RAF Marham, but then I realised he was dead right and I thanked him for his advice. Like a few other people have done during my life, Gary taught me a valuable lesson that I have always appreciated, and I have since that day never left anywhere in the RAF or future civilian employers without having a good relationship and that has served me well. So, for my last two months at RAF Marham, I kept quiet about Germany, RAF Bruggen and Jaguars; but inside I was bursting with excitement.

7. Bruggen and 'Real Jets'

RAF Bruggen started operations in 1953, and the base was named after the nearest rail depot. But it was located on the German-Dutch border in a village called Elmpt in North Rhine-Westphalia, which was about fifteen miles from the city of Monchengladbach. It remained a major RAF base until 2001, when it closed and, as usual, the British army gratefully took it over and renamed it Javelin Barracks. The army was booted out in 2015 and the base was used by the German government to accommodate asylum seekers and refugees and to become an energy and industrial complex.

It was in August 1981 when I flew into RAF Bruggen as a twenty-three-year-old lad. After doing the various induction training courses with the Rocks and settling in to a four-man room in the transit block, I was on my way back to the UK; this time, in north Scotland at RAF Lossiemouth, to do a two-week 'Q'—Jaguar-Weapons course to qualify me to work on Jaguar aircraft. But I went via Norfolk to see Sam and get my BMW to ride up to Scotland.

It was quite an epic journey and as I passed Gretna Green, I realised how much I had underestimated the time it would take me to ride the five hundred miles; I had to stop for a nap in a ditch near a lay-by. I had only been to Edinburgh to see the tattoo with the ATC when I was about fourteen years old and so this was really my first exposure to the beauty of the mountains and glens.

As I rode up through the highlands and turned into one glen at quite an altitude, the view suddenly revealed in front of me and literally made me gasp. I parked my motorbike on the side of the road and just took in the scenery for about ten minutes, simply staring in wonder at the unspoilt beauty of it. I had loved living in the mountains of mid-Wales for a couple of years as a boy, but the sheer epic scale of this landscape just blew me away. After the training course, I picked a different route back down south that came more westerly via Glencoe and again the staggering beauty of the landscape just amazed me.

However, when I stopped to take in the view this time, I was assaulted by the infamous Scottish midges and so had to move on quickly or get eaten alive!

I soon settled into life on 20 Squadron and quickly learned the ropes. It is said that the Sepecat Jaguar GR1, which was an Anglo-French venture, was designed as a jet trainer and so had maintenance and ease of repairs at the forefront of the designers' minds, rather than flight performance. The scope was soon changed, and it would become a supersonic aircraft with reconnaissance capabilities as well as a conventional and nuclear bomber role. It was a very reliable aircraft and the design used modular components that could be quickly exchanged on the flight line. Even an engine change could be carried out relatively quickly. The 30mm Aden guns could be reloaded or even replaced in minutes, likewise, the underwing external fuel tanks and wing pylons. In fact, the only armament jobs that were fiddly were the centreline pylon underneath the fuselage, and the canopy jacks that lifted the canopy into the slipstream on ejection needed dexterous fingers to do the piston mounting bolts. I found that having digits like pig's tits, was not helpful in this regard!

Later, they introduced tandem beams on the inner pylons so that additional two 1,000lb bombs or 600lb Cluster Bomb Units (CBU) could be carried. We did trials with just 1,000lb bombs and watched as the 'Pifco GR1 Jaguar' loaded with eight 1,000lb bombs struggled to get into the air. Despite almost using the entire runway, the pilot was pretty shaken up when he landed!

After that, obviously because somebody high up had decided we had these new tandem beams that created two store carrying positions out of one, we had to use them. The standard load in this configuration tended to be a centreline 1200 litre fuel tank, four 1,000lb bombs and a Phimat chaff pod (which throws out bits of foil to confuse missiles fired at the aircraft) and Westinghouse ECM pod (Electronic Counter Measure jamming equipment).

Years later, in the first gulf war, they also fitted over-wing sidewinder launchers to carry sidewinder air to air missiles, which must have been fun to load. The fantastic reliability of the Jaguar and ease of maintenance meant that unless we were doing routine maintenance such as removing pylons or ejection seats, our nightshifts, which started at 4:30pm, could be finished by 8:30pm; so, sometimes, we had just a four-hour working day. We even had an agreement between the shifts that if the dayshift could finish all the planned maintenance work before they knocked-off, then they would. Several times, I was telephoned at home to be told not to come in and just be on standby in case there was a problem. Years later, when I was on the brand-new GR5 and GR7 Harriers, where every snag and problem was a new experience for everyone, I would often not get home until 3:00am and how I missed my part-time job working on Jaguars!

I remember that we did work late one night when he had an unusual amount of work to do and I was told to lock up one of the HAS when I had finished working on one of the jets. I locked the heavy steel wicket door and as there was a shortcut back to our line hut through the woods, I went down that path. It was a pitch black, moonless night and I didn't even have a torch, but as I could see the lit area of the offices in the distance through the trees, it was not a problem. I was about halfway along the path going through the woods when I heard bushes rustling behind me, then I heard the snort as the animal started charging towards me. I now knew what a large pig sounded like, and I immediately knew that this was not a nice placid one like Tosh had; this was a large, wild boar with large tusks, and it was clearly not happy that I was on the path. I just took off and ran as fast as I could towards the lit buildings and I didn't stop until I was in the car park, bent over double, trying to get my breath under control. I don't know at what point the boar gave up the chase, but I think I must have covered the 300 metres or so in Olympic record time!

20 Squadron was special, a bit like armourers are a bit 'different' from the other trades. 20 Squadron had something about it, probably because of the people on it. I was working with Eddie Reilly, Tom Gibbons, Pete Fairley, Nev Carmichael, Nige Strawbridge, 'Oddball' (can't remember his proper name), Phil Jessiman, Dave Pepper, Tiny Walton, Dave Elwell, Ozzie Hoyte, Bob Ridley, Roy Ball, Dick Stinton and Mick Hayes. It was my first proper squadron and I just loved being part of it all. Only later, after being on several squadrons, would I appreciate just how good it had been.

By coincidence, my old roommate at Gutersloh, Jez had also been on his fitter's course. After a stint at RAF Wattisham in Suffolk, he had also been posted to Bruggen, but on 14 Squadron at around the same time. We met up a few times and when he was allocated a married quarter out at Wickrath, which was about twenty miles away, I agreed to give him a lift on my motorbike to take it over.

On the way, we rode through a large village called Niederkruchten and, as we approached a crossroads, the traffic lights were turning from green to amber, but I was already too close to stop so I just carried on. Suddenly, a vehicle pulled out from the right and was right in front of me. I probably braked, but it was all too late, and we slammed into the side of the car. The bike stopped abruptly, but Jez and I flew forward and summersaulted over the car onto the road on the other side. Amazingly, neither of us were hurt in any way, but obviously really shaken up. The German driver got out and seemed more concerned about the damage to his car than if we were hurt, he then started shouting at us. I explained that we were English and I that I only spoke a little German; this just infuriated him even more. Soon, the German police turned up and he did speak basic English, and it soon became apparent that the driver was claiming that we jumped the red light and so the accident was my fault. The officer had his notepad out and was starting to write his report when another gentleman, out walking his dog, came over and he started talking with the policeman. Apparently, he was on our side, and this made the driver incandescent with rage. Anyway, I was not blamed and eventually I got a settlement cheque from his insurance company, but now I had no wheels. The BMW was a write-off, and I only got a few hundred Deutschmarks for its scrap value. Even if I could have rebuilt the bike, I just did not have the heart to do so. I was also aware that I had survived, unharmed, two very serious accidents on that motorbike—perhaps it was a sign to leave it alone.

Fig 12. German VW car beats German Motorcycle in a crash – the end of my Beemer!

A few weeks later, I was on leave back in Norfolk and staying at Sam's, but not in the same bedroom. Tosh came up with the solution to my lack of transport from an advert from a private seller in the local newspaper selling a left-hand drive Mercedes 220D in a village near Norwich. We drove over there and there was this dark red huge Mercedes car but with a cream-coloured front wing and door, so it had obviously been in a crash. We checked it over and despite being ten years old, it was solid with no obvious rust spots anywhere; the engine was not covered in oil, though, it smoked a bit when it was first fired up and it drove wonderfully and was very smooth and comfortable.

It had done over 180,000kms, but Tosh pointed out that this meant it was just run in properly and, being German, would last for years. He was right as in 2017, a similar model was donated by a Peruvian taxi driver to the Mercedes Museum, and it had 3.5 million kms on the clock. I think I ended up paying about £230 for the car, which did have a few months MOT left to run, and drove it back to Tosh's place. Now, I should probably explain here that in rural Norfolk, there were not only honesty boxes outside many farms where you left money for any eggs or fruit and vegetable that was on sale on the roadside table, but indeed, the entire place had a bartering plus money economy. Tosh would get his car serviced in the village garage in exchange for a brace of pheasants or ducks or perhaps a choice piece of meat from whichever poor pig now resided in one of Rosie's several freezers. The only money that would be involved was for parts and oils that had to be purchased in that annoying modern era way that needed actual currency.

So, when Tosh had arranged to have the Mercedes completely resprayed, I should not have been surprised that it was only going to cost me £30, and by the way, did I like a dark red maroon colour rather than its current red? I did like the colour chosen and it's probably just as well, as I found out later that the maroon was the only colour on offer, as it was an unused job, probably from somewhere slightly dodgy. Anyway, a few days later, the freshly new painted car was delivered and it looked amazing. I drove it back to Bruggen where it became a favourite with the lads who named it 'the tank' as it was a big beast and about seven of us could squeeze in it at a push to go over to the mess for food from the squadron dispersal.

I had to go find some new suspension units from a scrapyard over the border in Holland for it to pass the German equivalent of the MOT, but then with German motor insurance, it was all sorted. I would drive back to see Sam in

Norfolk as often as I could, even just for weekends, and the tank made the journey at least ten times over the next six months; it was not very fast, but being a diesel, had a good torque and could sit at 135kph all day long. Booze was so cheap that it was more cost effective to use cheap Russian vodka as de-icer windscreen wash in the winter than purpose made screen wash.

I was driving to Calais to get the ferry to Dover for my Christmas leave and it was one of those nights where there was wet roads and freezing fog and so spray from other vehicles would cover the windscreen in muddy drops and then the water would start freezing; so I had to keep using the wipers and squirting the washers every twenty minutes or so. Soon after, crossing the Belgium-France border, I fell asleep at the wheel and awoke just in time to avoid running into the central pillar of a bridge. I parked the car in a state of shock, got out for some fresh air to wake myself up and to calm myself down again. When I went to get back in the car, shivering a few minutes later, as I opened the driver's door, the overpowering stench of cheap vodka flooded the air, and I realised that I was probably intoxicated and most likely currently a drunk driver!

Despite the freezing air, I had all the windows open for the rest of the journey to Calais and when I next refuelled the tank, I splashed out on some proper non-alcoholic screen wash.

Fig 13. Enter 'The Tank' which could carry 7 hungry armourers to supper on 20 Sqn night shift

When Sam moved over to Germany and got a job, we needed a second car and so I got a bank loan for a bright red second-hand Golf GTi, Mark One, the original German hot hatchback for about 8,500 Deutschmarks. In terms of a driving experience, it was the other extreme to the tank as this thing raced to 60mph in about nine seconds and had a top speed of 113mph, which, in those days, were impressive performance figures. It was like being back on my motorcycle but sat comfortably in a car. I loved that car and wish I still had it as they are a collector's item now.

Fig 14. The VW Golf GTi Mk 1 – did not realise at the time we owned a future icon car!

But before I had the GTi, the tank took me on a very significant trip back to Norfolk in March 1982 for my wedding at Sam's family's local church, in Shouldham near RAF Marham, with a reception planned in the village hall. On the day, it was very blustery with rain showers and with bitterly cold easterly winds that had everyone shivering in the hilltop church, let alone outside. I had stayed with Jez, my best man, at one of the Marham armourers houses for a bit of a stag doo and then I went straight to the church. I wasn't sure I was doing the right thing, but Jez insisted as "He had done it too, so I wasn't getting out of it, and anyway you can't be happy for all of your life!"

I had decided I didn't want to get married in uniform and so I had a suit on, of which I thoroughly checked the elbows for holes, naturally. Just before I walked into the church, some of my family from Torquay turned up and my step mum Sylvia gave me a hug and wished me luck, as did her mother and her sister, Brenda.

Then my dad came over and I was preparing myself for some sort of council or worldly advice, but instead he informed me that they would not be at the reception as they had a long way to travel back to Devon; this was despite us already having paid for them to stay in a pub B&B down the road. I pointed this out, but no, he and Sylvia were off after the church service. Apparently, he had been very cool towards Rosie and Tosh in their house earlier and made everyone feel uncomfortable. Sylvia came to me and said she was so sorry that my dad was ruining the day, but she could not make him change his mind.

So, into the church I went and tried to just concentrate on what was about to happen. Sam foolishly decided that she would rock up after all and the service went without a hitch. The wedding photographs taken outside the church afterwards show everyone apparently smiling, when in reality, they were grimacing at the wind, rain and cold and trying not to shiver. After the wedding ceremony, everyone thankfully dived into their cars and, no doubt, put the heaters on full blast as they made their way to the reception at a village hall. Everyone piled in ahead of us and made a lovely guard of honour as we walked in.

My parents had left by now, and Sylvia's sister (my aunt), Brenda, came over to us and said, "Your father is a piece of shit."

However, we were not to worry as her parents Cissie and Bert were going to sit in as my parents on the top table. Sam was so relieved and gave her a big hug and said thank you. I had never heard Cissie, a woman of Scottish heritage, swear in all the time I had known her, but she aired her views on my father's actions in words that turned the air blue, she was so furious.

The reception went well and soon the meal turned into a drinking session until the evening doo started, when lots of my armourer mates from Marham came along. The party got going and the dancing went on long into the night and so, despite my father's unfathomable actions, everyone had a great time. I was now also part of a normal loving family that actually cared about me as well as each other.

We then had a great week on honeymoon in Malta and, all too soon, I was leaving Sam in Norfolk and driving back to Germany. The wedding episode sadly created a schism in the family that never healed, especially between the sisters Sylvia and Brenda, which was so sad. Brenda was a wonderful person; when I was in basic training at Swinderby, I was sent an anonymous valentine's card from her that she had sent just to make sure that I had one. It was a gesture that I really appreciated, despite the initial disappointment when I realised that I didn't have a secret admirer somewhere. She survived cancer in her late sixties but then had a fall in early 2022 and died from brain damage a few days later. I went down to the funeral in Torquay, and it was so sad that the world had just lost such a caring person.

When I got back to work on 20 Squadron, the Falklands conflict had started, and when we started losing ships and people, I remember feeling so angry and helpless and I put in a General Application to go fight – not that my presence would have added anything useful and it was declined! One Sunday afternoon, a couple of us decided to go out for a drive and we ended up in Arnhem; we went for a beer near the infamous bridge of the World War II film, A Bridge Too Far. An elderly man was sat at the bar, and he started buying us schnapps. When we tried to buy him a drink back, he almost got angry and explained that the drinks were for our fathers and uncles that had fought in the war against the Germans.

When I explained that our fathers were still boys when the war started, he was unperturbed and said, "Well, for their fathers then, and all British fathers that helped us against Hitler."

We had no answer to that, so we just thanked him for the drinks before heading back to camp. I lived in the transit block for another few months and eventually, I was allocated a married quarter on the top floor of a block of flats in Wickrath, just across the road from where Jez and Jos lived. Soon, Sam was with me and we were setting up our home. In typical forces lifestyle, as soon as she arrived, I was off on the annual Air Practice Camp (APC) detachment to Decimomannu, near Cagliari in Sardinia. The Jaguar squadrons always got the summer slots and so, our weekends were spent exploring the island or going on the RAF bus down into town or out to the Fortis hotel complex, where we would invade their beach and try and sneak into their pool areas. Once in Cagliari, we even had the brass neck to go into one of the dining areas of a tourist hotel, give a false room number and then treated ourselves to a free meal and drinks. We got caught out once and thrown out of the hotel when we said a room number on a floor that didn't exist and you would think that was fair enough, but later that night, around 1:00am, we returned and set up the bottom of the pool with sun lounges, chairs, a table and even a large sun umbrella fully opened under the water. God knows why we did this, especially as we would not be there in the morning to see the shocked faces of the pool area staff, but that is the power of booze, I guess! I went to 'Deci', as we called it, several times with three different squadrons, so I don't recall exactly which year the next event took place, but I think that the tale is worth telling.

The Americans came to Decimomannu for a brief visit, and they brought some Grumman F14 Tomcats in from one of their aircraft carriers sailing in the Mediterranean Sea. The pilots announced their arrival by doing a low fly-past and then, one of the F14's did a solo display out over the airfield; it pulled up vertically at about two hundred feet above the ground and, just using the sheer power of its afterburners, sat vertically in the sky on its tail and gradually went around in a circle, bobbing up and down to maintain its position. It was one of the most impressive bits of flying I have ever seen. I think such a manoeuvre is banned now so I'm glad I witnessed it back then.

Whilst the aviators were impressive, their support ground crew were not so much. These US Navy guys came into our bar, full of themselves and the usual bonhomie that naturally exists between forces personnel of all friendly nations just wasn't there. These guys were here to make sure that we all knew that they were the best. Drinks flowed but the patio area was divided between us and them. They were loudly proclaiming that they had the greatest fighter plane ever built and "wasn't our 'Jagwarh' just a trainer aircraft?" They were right, of course, and to me it felt like playing for Scunthorpe United against Real Madrid! So, we were trying to ignore them when they decided to make things interesting. They challenged us to a singing contest and then delivered a pretty good rendition of 'God Bless America' for a couple of verses. Although this was a squadron gathering and we had squadron songs, it was naturally the armourers that let rip immediately with the immortal armourer's song that explains that we like to rape old age pensioners, steal children's toys and have sex with Gorillas. Understandably, our American cousins looked bemused.

Then, in really bad taste, someone started the moron joke song, which starts with a spokesperson shouting out, "What does a moron sound like?"

And everyone goes "Dur."

This is followed by two morons getting 'dur, dur' and then three morons getting 'dur, dur, dur'.

Then comes the punchline, "What does a whole nation of morons sound like?"

To which, to the tune of the 'Star Bangled Banner', a response is sung with, "Dur, dur, dur, dur, dur, duuur, dur, dur, dur, dur, dur, duuur."

It went quiet for a moment as the US Navy guys processed this and then, correctly decided that we had just collectively insulted them, their nation and their national anthem; they were not happy about it and were ready for a fight. But you could also see that they realised they were playing away from home and heavily outnumbered.

So, there was an awkward moment, which was very neatly defused by their chief, who shouted out, "We representatives of the US navy would like to buy all of our British friends here a beer."

That guy clearly had a potential career in politics ahead of him as it worked, and the situation was totally defused, and an insult-free party duly kicked off. There was one more highlight from that evening for me, when I was chatting with one of their safety equipment guys (or 'Squipers' as we would call them). He was bragging about how if a F14 pilot ejected, then their survival packs had a gun, knife, gold sovereigns, ration packs and bottled water. Our survival packs had a penknife, but I didn't tell him that; instead, I told him as a joke, that we also had powdered water in our packs to reduce the weight.

To my amazement, he said, "Really? Wow, how does that work?"

And so, I said, "You just add water to the powder."

And I was astonished when he said, "Wow, we don't have any of that stuff. Can I come and get some from you tomorrow?"

I told him that he could and I tried to keep a straight face! This all took place before we knew the truth about the Falklands War and which, without US tacit support, would probably have been a failure for the UK. Years later, when I was working with the US forces bomb disposal team during the first gulf war, I was so grateful for their support. I have a much greater respect for the US military and Americans in general now than I did back in 1982! Years later, I would go with a Harrier squadron in rainy and cool December to Decimomannu, and it would be a very different experience to my summer sun-soaked Jaguar detachments.

As the film 'ET' was released in the UK, we were discovering that as newlyweds without kids, our life was great. We had a busy social life and we would do a lot of things with the single squadron guys, including a long weekend in Munich. The 'singlies' also had a habit of enforcing the 'Taceval beer rule', which meant that as long as they brought a slab of beer with them, any singly could knock on the door of any married guys, known as 'Scaleys', which may have derived from married servicemen being on Scale E of payments. The Taceval beer rule meant that in return for the beer, the singlies could eat, drink and stay the night. Not being stupid, they targeted couples without children! I recall one time, we saw off one guy after a long night, and another two rocked up. Fortunately, as we lived a long way from the base, this did not happen very often and, over time, our social life switched from the single lads to other couples in our married quarter area, including Jez and Jos.

In the autumn, we did the wine festivals on the Mosel, and the 'Rhein in Flames' festivals at Koblenz; I can still hear the German oompah music and the tantalising smell of grilled meat coming from the swinging steaks over the BBQ. I have less fond memories of the monstrous hangovers that followed such wine drinking events. As the winter months drew in, Jez and I would go into the local town to watch Monchengladbach home matches and frequent the so called 'Street of a thousand bars', which was a misnomer as there were only about ten bars and several restaurants.

At the time, everyone was playing Michael Jackson's iconic 'Thriller' album and that Christmas, a bunch of us went into town dressed as Father Christmas, dishing out sweets to the kids in the area, before going to our favourite of the street's bars. A few hours later some very drunken representatives of Father Christmas emerged and hopefully, not many children were about to witness a jovial but not exactly model Santa representative! The Germans have also invented another winter carnival event in February on Shrove Monday each year to brighten the gloom—Rose Montag, which may mean rose (or derived from the word roose, which means frolic). Families dress up in costumes and take to the streets to have a party before Lent kicks in. Naturally, being British, Shrove Tuesday the next day just meant having pancakes to us; the Germans preferred to stay out drinking beer and Bratwurst and we all agreed that they had the right idea!

Getting married had made me realise that if we were going to be comfortable and be able to afford to raise a family, then, in addition to Sam's salary being added to my JT wage, I would need to get promoted. I was quite content as a JT as I had the technical work to do, such as installing and removing ejection seats, which was better than the mundane stuff I did as an SAC and I was paid only a little bit less than a Corporal, but without any of the responsibilities of having rank. In order to be considered for promotion I had to pass a written 'Weapon's

Trade Test', which was held once or twice a year. I applied to take the test and soon had all of the material needed to learn about explosive regulations, and so I started revising for it.

To be fair, Sam probably knew as much about these regulations as I did because every night, before we went to sleep, she tested me with the fifty or so questions that I had developed from reading the material and the practice exam examples provided. There was a mock exam and I scored 97% and I found out that I had passed the actual exam with 'distinction' a few weeks later. Years later, they did away with the exam, which I thought was a mistake as you had to be motivated to pass it and so the people that got promoted really wanted it.

Fig 15. XX Squadron all ranks (circa 1982)

Early in 1983, back on 20 Squadron, the armament Chief Eddie Reilly called me in to his office. I remember Eddie as not only being a good boss, but also because in his married quarter, he had a 28-inch television, which, at the time, was huge. I remember commenting on its size and he justified it as the TV was their main source of entertainment, which made absolute sense to me. As I sit here writing this, I can see our massive black 80-inch TV and I just love watching sport and movies on it, and when people comment on its size, I use Eddie's justification line. Having called me in, Eddie told me I had been selected to go on the squadron Red Flag exercise detachment at Las Vegas that summer and I was made up. This was the detachment that everyone wanted to go on, so could my life get any better?

Apparently, it could, as I was also promoted to Corporal and packed off to RAF Hereford to complete my General Service Training (GST) course, which, as I recall, was mainly about service writing and fieldcraft soldier type training. I can only really remember one funny story from the training, which was when we were being taught about defensive gun placements to create a defensive covering fire. Each gun pit has a range card with arcs of fire using distinctive objects and distance markers so that you could set your weapons to the correct range if an enemy was advancing on you. On the card, in the gun pit that I took over, the previous occupants had developed the range card and, as one reference point, they had put 'herd of cows'. Presumably, they were city folk who didn't realise that cows move around the field and go for milking twice a day! I completed the GST training just after my first wedding anniversary and I now had two Corporal stripes on my arm.

Fig 16. Me, front row, 2nd left on General Service Training

A few weeks later, I was sat in a VC10 transport plane next to Tom Gibbons, and we landed first at Washington for a refuel and then at Nellis AFB, Las Vegas. To make this dream trip complete, we were staying in a casino called the Bingo

Palace and we were handed a wad of US Dollars as spending money. We would have to put up with five weeks of this; you know, sometimes life in the RAF could be so hard!

Las Vegas is a very exciting place for a twenty-four-year-old lad who was given free accommodation and a wad of extra money to spend there each week; and, when not at work, it was simply an amazing experience. On the first night, we left the Bingo Palace and headed across some wasteland to the main strip and the nearest big casino called Circus Circus, which, as the name implies, was a casino with massive rooms of one-armed bandit slot machines, casino areas all themed as a circus but with additional fairground attractions. The British lads were soon banned from the darts game in which you had to throw darts at playing cards to win a prize; they found it too easy and soon all had huge teddy bears and their other top row prizes.

Later, I was stood outside Caesar's Palace, where an illuminated billboard announced that Tom Jones was the resident act and inside, I was wowed by the blue dome roof that made it look like it was daytime in some roman city. Then, we got a taxi down to the oldest part of Vegas—where the huge cowboy stands that appeared in so many films over the years. We were all hyper excited and as we had been given a late start the following day, the night was a long one and some of the guys never even made it to bed!

So that this doesn't just look like we had a nice holiday in Las Vegas at the UK taxpayer's expense, I should point out that we did actually work on the same sort of day/evening shift pattern that we had at Bruggen. The Red Flag exercise started and, as expected, our boys got trounced by the superior US aircraft. They did win one 'battle' when our aircrew got a bit sneaky and sent off two Jaguars as a decoy to attract the US pilots who fell for it. That left the other Jaguar to fly low, probably illegally low, to fly in and hit the target unhindered. But other than that dubious victory, our pilots suffered a right pasting. But of course, we did not care one iota; we were simply working when we needed to and having the time of our lives when we were off, and every week, they kept giving us more spending money.

On our first long weekend, as we finished Friday lunchtime and would not be back to work until Monday evening, we went to Los Angeles. After a really boring car journey, we did the Hollywood tourist sights, Rodeo Drive, The Walk of Fame, and the original Disneyland and Universal Studios tour. The following normal weekend, we drove to the Grand Canyon and stayed the night after really livening up the local bar with our amazing singing.

On the way back, I was driving the truck with six of us in the two rows of seats. I was keeping diligently to the 55mph speed limit but everyone was keen to get back to Vegas and so they were badgering me to speed up. In the end, a pact was agreed that if we got fined, then all six of us would share the cost of paying for it. So now, I booted down the accelerator and we were soon flying back towards Vegas with everyone's eyes peeled on the lookout for traffic police cars. It was getting dark and we were making good time, when someone spotted a police car and shouted out a warning. The police were not parked on the side of the road where we were all looking but parked in the central reservation and the traffic cop was stood next to his car with his radar gun pointing right at us. I slammed on the brakes, but it was all too late and sure enough, a few minutes later, he was right behind me with his blues and twos going and pointing for me to pull over.

What happened next was a bit like the 1970s Jasper Carrott sketch, where the American policeman assumed that as Britain is so small compared to the US, we must all know each other, and he asked Carrott if he knew a Mrs Harris from Blackpool. The policeman walked up towards the car from the rear and had his hand on the pistol of his gun, which is actually quite unnerving for an unarmed Brit sat in a car.

I wound down the window and the policeman looked at me with an exasperated, '*when will they learn?*' face on' and asked me, "Evening, sir. You in a hurry? Do you know how fast you were travelling back there?"

I put my best British posh accent and said something like, "I'm so sorry officer, I didn't realise how fast I was going until I saw you and then checked my speedo."

To which, he replied, "Well, you were clocked doing 83mph and that was with your brake lights on, and all four wheels locked up." Followed by, "You do know that the speed limit is 55mph, right?" and then, "Let me see your driving licence and vehicle ownership papers."

I handed over my UK driving licence and explained that the vehicle belonged to the US Airforce, and we had permission to use it for a recreational trip to the Grand Canyon. I also explained that we were actually based in Germany where there were no speed limits on the freeways as if that was a good excuse for speeding. He nodded a bit and considered the situation, and I was thinking we would be getting locked up in some country jail and be in deep shit when we got back to base.

But to my surprise, he actually said, "Listen, I am all for our armed forces and I love you Brits, but you can't abuse our traffic laws without punishment. But what I'm gonna do is pretend that I have pulled you over five miles further up the road in the next county as they are not as hard on enforcement as where you are right now. I am also gonna state that you were doing 69mph so that you avoid a court hearing."

We all obviously thanked him profusely and gladly accepted the lesser punishment. The next day, I went to a sort of post office and paid the fine and I think it cost us about $19 each.

Being on a deployment for five weeks in a holiday venue is very different to actually being on holiday; in that, work is the main part of the daily rhythm and after the first few giddy and breathless days of going out into town as soon as we got back from work, you settle down into a routine. It was strange to be actually living in the Bingo Palace as it was a small casino; and, to go for our breakfast, which was paid for, we had to walk through the casino itself in our uniform past all of the gambling tourists.

The main memory that I have of Vegas is the noise in the casinos, which all look very different on the outside, but which inside are basically the same. Casino areas with the traditional gambling games of cards such as poker and blackjack, as well as roulette and the unfortunately named dice throwing game called 'Craps'. But these are played to a background cacophony of noise created by the massed ranks of one arm bandit slot machines as they operated with sounds and music and frequently spewed out coins, clanging into the metal coin trays as winnings. Occasionally, a loud bell would ring if someone had been lucky enough to win a big prize and over the speakers, numbers were being read out for the game of Keno, which is like Bingo. Add to that the general hubbub of people, cheers and groans from the casino game area and cocktail waitresses going around delivering drinks and collecting orders for the free drinks you got whilst gambling, and you realise that everything is designed to create a toxifying atmosphere of excitement and air of risk and reward. There are no clocks to remind you of the time or make you realise how long you have been gambling for in this entranced state.

Years later, I flew in on holiday with Sam to the civilian airport and was amazed to see slot machines all over the place as we arrived! We soon realised that the best way to live cheaply in Vegas on our allowances was to eat fast food in the Dennys, McDonalds, Burger King, Pizza Hut nearby or go to the cheap buffets offered in the big casinos that were designed to lure you in to eat, before you then got gambling. A more expensive option was to go to the less popular shows that had a dinner option as part of the package, and I went to see George Benson, as much for the good meal as his music.

The shows in Vegas are amazing and during that holiday in 1995, we went to see the famous Siegfried and Roy show at the Mirage, before 2003 when one of the tigers called Mantacore decided he was done with all the performing and attacked Roy. They were great illusionists and to this day, I don't know how they made an elephant vanish so quickly in a box, and then reappear slowly up under a sheet onto a stage that I could see clearly right underneath, and now, of course, I don't want to know and spoil that magic.

Most tourists go to Vegas just for a few days and they make the most of it. I can clearly remember going for early lunch on a Friday, as shift change was at 12:30pm for the night shift, and I noticed a middle-aged woman playing three slot machines at the same time and she had a bucket to contain the coins that she was feeding to each machine, four coins at a time. I don't know why I noticed her, but at 10:00pm, when we got back from work and we were going to do our usual money saving trick of playing blackjack with one-dollar bills for a couple of hours whilst drinking the free Harvey Wallbangers heavily, she was still there at the same three machines.

Next morning, when I came down at around 10:00am for breakfast, I was shocked to see her sitting there again. I thought she must have gone to bed, but she appeared to be in the same clothes as the day before; so, to my knowledge, she had been sat there, feeding those three machines for twenty-three hours. At least now she had two other buckets full of coins, and so I guess it was worth the time and effort to her.

Soon, it was our final weekend and a few of us visited the impressive Hoover Dam, did the guided tour and then visited Lake Mead, which the dam created. I was amazed that if you threw a bit of food into the water, it would suddenly boil up with a mass of squirming Carp fish trying to eat it. When the droughts hit in 2021, I noticed how low the lake water levels are and I wondered how the fish were doing with much less living space. Then it was time for the long flight home on a VC10, which, because we were not as excited as we were on the way out to

Vegas, seemed twice as long, especially because after five weeks, we were keen to get home to our families.

When I got back to Bruggen, I was called in to see the Junior Engineering Officer (JENGO) and he informed me that whilst I had been away, I had been posted to 31 Squadron down the road. I was really sad to leave 20 Squadron as I had really enjoyed myself there. Dick Stinton recently described it to me like living in the Delta Fraternity from the film 'Animal House', and he was right. However, I was also relieved that as 31 Squadron were also Jaguars and based at Bruggen, there would be no house moves or training courses needed. Not long after I arrived on 31 Squadron, I found myself on detachment to Decimomannu and I soon settled in, but this squadron had a very different feel to it.

On 20 Squadron, the aircrew and officers joined in with the lads more, especially on detachment; but here, on 31 Squadron, they kept their distance more. Another difference was how the armourers were treated by the other trades, especially the Fairy avionic guys. On 20 Squadron, there was a united ground crew that did everything together, but here on 31 Squadron, the armourers were treated differently, especially by the fairies—even being mocked for not being very bright. It was the first time that I heard the song "'A': I'm an Armourer, 'B': I'm an Armourer, 'C': I'm an armourer," despite having been in the RAF for five years. In response, the armourers pointed out to the fairies that you didn't have to be clever to just change avionic boxes on the Jaguar, which were then sent to a servicing bay on the base for repair, and that this was probably why armourers were paid more at each rank than the fairies. At the time, I thought it was an amusing retort, but I didn't like the split in the ground crew and some of our own armament SNCOs seemed to like keeping the disharmony going. Although the squadron had a different feel to it, naturally, the armament team were a great bunch and I had many fun times with the likes of Martin Davis, Mac McGregor, Billy Anderson, Gordon Hunt, Ollie Holland, Alec McCreadie, Steve Darrall and Les Waite.

On a detachment to Deci in Sardinia, one of the fairies was a JT single girl who preferred to knock about socially with the other trades and, understandably, the armourers. But her chief technician took exception to this and was not happy about this at all as he liked all of the fairies to do everything together with him as the leader, so on detachment, they all went to the bars and restaurants. She ignored him and came out with us to the German bar one night and he came in like a possessive father and dragged her out. I went outside to remonstrate with the chief, but he was furious and threatened to 'punch my lights out' if I didn't go back into the bar. As he was a big guy, I bravely did just that. She was crying by now and then the rest of the fairy gang turned up and they all went into the bar.

Inside the bar, the chief came over to me and I thought he was going to apologise, but instead he told me that he was going to get me before the night was out. I got another beer and sat back down with the armourers, but I didn't tell them everything as I still thought the chief was just having a bad day. SNCOs don't generally go around threatening to beat up their junior ranks, and so there was no need to escalate the situation into an armourer's vs fairy battle.

But a bit later, I noticed the chief looking at me and pointing at his watch and then nodding to go outside; so, this was not going away then! When I had been outside earlier, I had noticed a piece of 4" × 2" wood lying next to the bar patio area and so I left the German bar and quickly grabbed it. Through the window, I saw that the chief had watched me go and he casually finished the last of his pint and headed towards the door. I thought I would only have one chance at taking this big guy down and so I stood to the side of the entrance door. As it opened inwards and a shape started to emerge, I swung my weapon with quite a bit of force at knee level, but I misjudged things badly as I caught a pillar that was holding a porch roof up. It was lucky that I did miss as it was not the chief but somebody just leaving the bar who was now jumping backwards, shouting expletives and wondering what the hell was going on. However, another of the fairies was right behind him and realised who my intended target had been.

Whilst I had been getting all Rambo-like, apparently, the chief had just gone to the toilet for a pee and when he heard about what had happened, he looked at me with a surprised expression, but didn't do or say anything. I don't know for sure in this case, but I suspect that like most bullies, when they are challenged, they show themselves to be cowards. Nothing else was said by the chief and nothing happened after that, though the fairy JT girl stayed away from the armourers and just went around with her own box changing fairy trade. I suspect that she probably either left the RAF early or died of boredom after that trip.

There was also an issue brewing with the pilot's mascot, which was a large teddy bear, a Panda, I think, dressed up in a little flying suit to look like a pilot and I think it was called Flying Officer Ted. The aircrew insisted on taking it away with them everywhere and the most junior pilot was responsible for its safety. Once, it was left behind accidentally in Bardufoss in Norway, probably because the detached squadron ground crew had hidden it. When the aircrew realised, they arranged for the twin seat trainer Jaguar aircraft to fly up there on a 'training mission' to retrieve it and the pilot put his skiing equipment into the baggage pod under the aircraft so he could maximise the trip. Sure enough, a couple of days later, the T-Bird returned with the mascot strapped into that back seat of the trainer. But that soft cuddly toy mascot would cause a massive issue a few months later.

As 1983 came to an end, I was called into see the JENGO and I was told that as one of the armament corporals, Paul Brindley, I think it was, had broken his leg, and that I would be going on the next Red Flag detachment to Vegas in his place. I knew that this would not go down well with the other armament corporals on the squadron, who had all been there longer than me and, of course, I had just come back from Vegas a few months earlier. The JENGO said that he was well aware of all this, but that I was the one that they wanted to take. Well, what could I do? Be grateful and go, obviously!

There was some resentment for a while amongst the lads, not all directed at me, but at the people who had decided to take the new boy over one of their own. As a result, just after the start of 1984, I found myself back in Vegas again. Unlike all of the others that were excited, as I had been a year earlier, I found that this time I settled very quickly into the work routine. Instead of just frequenting the casinos along 'the strip' again, I found myself seeking out backstreet places and somehow ended up in the Troubadour's, which was walkable from the Bingo Palace. The venue often had well known groups playing as well as up-and-coming acts, but these were not high profile or well-advertised events. They felt more like private jamming sessions for the locals, and this became the regular haunt for a few of the guys that were over the Vegas glitz.

The standout memory for me on this trip happened after a meal in a little Italian restaurant near the Bingo Palace. It was a cheap place decorated with plastic vines and, unsurprisingly, pictures of Italy. The four of us enjoyed a good meal and lots of wine and we had been noticed by one of the local girls, who was a very attractive blonde with a good sense of humour. Ever since the 1978 film Animal House, the chant of 'toga, toga' had often resulted in RAF personnel immediately dressing up in bedsheets and pretending to be romans. The topic of going to Caesar's Palace in a toga came up, as it had with 20 Squadron on the previous trip, but this time we had, let's call her Patty, egging us on and offering to give us a lift in her convertible car. Naturally, being unable to resist a dare, we had soon stolen some of the artificial vines from the restaurant and wearing nothing but our boxers and bedsheets from the Bingo Place, we were soon cruising down the strip in an open-top, leather-seated, red Mustang convertible car to the famous Caesar's Palace. The entrance to Caesar's Palace is a one-way system in and out from the strip with a roundabout at the top and a series of fountains and ponds with conifer trees down the central reservation. We drove up to the entrance and Patty dropped us off and where she went after that, I have no idea.

As we walked up the steps into the casino, I heard an American say, "Well, it's either a frat party or they are Brits."

No sooner had we entered, when a big burly security guard came over, who had obviously dealt with this sort of thing before, as he said, "Okay guys, you have five minutes on the dance floor over there and then leave. Got it?"

We agreed and I was soon on the dancefloor with some American middle-aged lady who just loved the fact that I was English. Of course, this was long before phone cameras and so, sadly or perhaps gratefully, there are no records of this momentous occasion other than long-gone security camera footage. The burly security guy caught my eye and gave the 'time to go' eject symbol with his thumb and I grabbed the other guys. We left to applause from the people around the dance floor and so, with a little bow to our new fans, we left. Outside, there was no Patty and no car waiting and so, we just hung about outside the entrance, wondering what to do. I was desperate for a drink as the magic power of alcohol fuelled exhibitionism was starting to wear off for me.

Then, one of the guys—and I'm pretty certain it was a rigger or engines guy called Wobbly Gob, earned his nickname because he seemed to have a can or glass of Warsteiner permanently at his lips—had decided that the fountain pool looked inviting. He jumped in, started splashing around and making whooping noises. The other two guys I was with thought this to be a great idea and jumped in too. By now, my sensible head had kicked in and I realised that I had to be the responsible adult – and it's not often that this happens! Soon, burly security guy was back with a mate, and he was not happy at all. Clearly, we had exceeded the toga acceptability threshold for Caesar's Palace and he shouted for the guys to get out, which they completely ignored. By now, the crowd of the casino's clientele, most of whom had followed us out, had grown even bigger. Hearing the security guard shouting, they just cheered us on and clapped at the antics of three drunken Brits in the fountain. Burly security guy was on his radio and getting pissed off that the crowd were making the situation worse. About five minutes had elapsed and I was still trying to convince the guys that they should get out of the fountains, when some remarkable things happened.

First, I heard a police car with its sirens wailing in the distance, getting closer with its red lights flashing, which turned into the long driveway of Caesar's Palace. Then, from out of a turning from the car park on the exit road, going the wrong way came a red Mustang convertible with Patty driving, her long blonde hair blowing in the breeze—what was even more impressive was that she then performed a perfect handbrake turn and the Mustang came to a halt next to us, facing the exit the right way now. *Who is this amazing woman that could drive like that?*

To me, it felt like the police car was getting close now and the guys in the fountain had now realised we were in a dodgy situation. They finally climbed out of the fountain. Wobbly Gob was dithering about near the edge of the fountain and I just grabbed him and pushed him into the passenger side of the car. Patty screamed at me to get in and so I just dived over the back door and had my head in the back seat floor, with my legs up in the air, presumably showing the world my hopefully clean boxer shorts (for the record, I now wear trendy Under Armour pants in case I wind up in a similar predicament).

To a mix of screaming tyres and cheering people, we hurtled off down the exit road of Caesar's Palace with a sense of relief and giddy excitement. The guys said that the crowd sensed that we may be able to escape, so they suddenly rushed forward and blocked the road in front of and behind the now trapped police car and its confused law enforcement occupants. I don't know if that was true or not, as I still had my head in the rear footwell. But I did think that the whole police car thing with blues and twos going, and armed officers was a bit of an overreaction to some pissed-off Brits wearing soaking wet bedsheets. Patty drove us back to the Bingo Palace via a circuitous route, to avoid any police cars that may be looking for us and dropped us off. We climbed out and looked in awe at the Mustang driving rock goddess that was Patty, but there was no talk of seeing us again, or future drinks.

She simply said, "Thank you, guys, that was such fun. You rock!"

And then, without another word, she booted the accelerator, the tyres squealed as they grappled for traction, and she drove off into the night. We never saw her again. For me, that just added to her aura and I hope that she continued to live her life in that exciting and carefree way and not end up living in some dreary suburb with three kids as she was the epitome of 'Girl Power' long before the Spice Girls came along and rocked the planet. The next day, I was called into see the WO Engineering and he asked me if I had been anywhere near Caesar's Palace last night.

I hesitated, thinking about what to say, when he said, "Corporal James, the answer you are looking for is no."

So, taking his lead, I simply lied and said, "No."

To which he said, "Great, I can tell our hosts that it wasn't us, but I suggest that you and your friends stay away from the strip for a while," and he winked.

The rest of the detachment passed without incident, and it was time to head home to Germany. The VC10 that was flying out was making a stopover for forty-eight hours before flying us back to the UK, presumably to do with flying hours that would also let the aircrew to rest and take in a show or something. There were four spare seats on the flight that would also not be needed on the way back and so the WO Engineering had arranged for four of the squadron wives to come out for a weekend in Vegas. But when it landed, two wives of the pilots got off and only one wife from the ground crew, as the wife of one of the engine bashers had not made the trip. Then, we found out why.

It was Pilot Officer Ted, the pilots' teddy bear mascot, that had been given the seat instead. As you can imagine, sacrificing someone to have a free trip of a lifetime so that a stuffed toy could be seated on the plane did not go down well. This meant war and there was no way teddy was going back on that plane! Several attempts were made to kidnap Pilot Officer Ted, but the aircrew were doing a pretty good job of keeping it secure and they almost made it. On the apron next to the VC10, we were waiting to board after the last cargo box had been loaded on. One of junior pilots was guarding teddy when one of the airport staff came out of an office and shouted his name and gave the telephone call to him. Surprised, he walked over to the office and went inside, then a few seconds later, burst out again, realising it had been a diversionary tactic, but it was too late. Pilot Officer Ted had gone. Fearful of the consequences of his guarding failure, he rushed about, looking for the teddy and gave us pleading looks for help, but nobody was in the mood to save teddy.

In the few seconds the pilot had been gone, the teddy had been taken, stuffed into a vehicle and driven over to the canteen on the side of the apron. Inside, the lady member of staff looked on in amazement as the teddy was stuffed into a chest freezer and told that after the white transport plane had gone, she could consider it a gift to her children. Soon, it was time to board, and around ten of our pilots were hunting around the apron and hanger, desperately looking for the teddy.

One of them went into the canteen and we held our breath, but a few seconds later, he came out, empty handed. It was job done and teddy was staying in Vegas. As the VC10 rumbled down the runway, a rendition of the Middle of the Road song 'Chirpy, Chirpy Cheep' started up at the back of the plane. *Where's your teddy gone? Where's your teddy gone? Far, far away.* Naturally, there was a price to pay for this heinous act and there were no monthly squadron beer calls and the relations between aircrew and ground crew were really quite unpleasant for a while. Eventually, Wing Commander Bolton relented, and a joint beer call was held.

During this, after a few beers, a large tri-wall cardboard box addressed to the Squadron Commander was brought into the crew room and the boss said, "I bet I know what this is."

He opened the box, expecting to find Pilot Officer Ted, only to find it full of flow pack polystyrene pieces with a postcard from Las Vegas sat on the top. After realising teddy was not in the box, he read the postcard to himself and then made a great effort of giving a smile of appreciating the joke, but we could all see the redness in his face and the throbbing blood vessels on the side of his forehead as he soon left.

The WO Engineering walked over to the box and picked up the postcard and said, "For f*ck sake lads, we are supposed to be mending fences here," and walked out hurriedly after the boss. Of course, we were all curious to know what the postcard said, which was *Hi Boss, having a great time in sunny Vegas, so have decided to stay here—lots of love Pilot Officer Ted xxx.*

During this time of aircrew vs ground crew disharmony, I happened to be on Quick Reaction Alert (QRA) duty. There was an area of Hardened Aircraft Shelters (HAS) adjacent to 20 Squadron that was specifically to keep a Jaguar on permanent standby, loaded with a WE.177C nuclear bomb with another aircraft that could be loaded within an hour, also housed in one of the other QRA HAS. The place was well guarded with double fences and patrol dogs at night.

The RAF Police sat up in high concrete towers, overlooking the entire area and they must have been bored shitless. When I was on 20 Squadron, we occasionally had to change the QRA aircraft over and so the gates to our squadron dispersals were opened. This always made the RAF Police in towers edgy, and we liked to make them feel better by shouting up at them 'Pigs in Space', which was a skit from the 'Muppet Show' at the time. This would usually result in getting two fingers shown back in our direction! As the nuclear loaded Jaguar had to be ready to go at ten minutes notice, we had to have a crew of a pilot and two ground crew living in the same compound 24/7 for a week.

I used to have a recurring bad dream of turning up for work at 20 Squadron and seeing the QRA gates hanging open and then watching the QRA Jaguar taking off from the runway and turning north-eastwards towards Moscow. We knew if that ever happened, we probably had just a few more minutes to live before Soviet nuclear ballistic missiles hit Bruggen and the other RAF and USAAF bases. Considering the security and care surrounding the nuclear weapons, it was hard to believe the rumour that we heard the following summer, which was that one of the nuclear bombs had actually fallen from a special S-Type trolley during transit. I have never seen any sort of bomb fall from one of the trolleys as they are normally strapped on securely, so to think that a nuclear bomb would be moved an inch without being strapped on was hard to believe. But years later, in 2007, the MOD confirmed that there had been a nuclear incident at RAF Bruggen due to the weapon not being secured to the trolley, I am glad I was not involved in that fiasco!

On the QRA, every morning, the aircraft had to have a 'Before Flight' (BF) inspection and service and the pilot would accompany the Corporal team leader and flight line mechanic (FLM) to the HAS to service the plane. The pilot would sit in the cockpit going through their drills. I imagine that the pilot's nightmares were about flying the plane on that mission, knowing that they may kill thousands of people in the unlikely event they made it to their city target and that it was definitely a one-way trip without a happy ending. As we entered the HAS, the pilot and FLM were through the heavy steel wicket door when there was a gust of wind, which blew the door shut with a bang. I opened it straight away and went in.

The pilot got straight into my face, shouting, "Two-man principle."

This was a rule—nobody was ever left alone when near the aircraft to reduce the chances of sabotage. Perhaps, rather stupidly, I pointed out that there had been two of them inside the HAS when the door closed momentarily, which was a valid point that the pilot did not appreciate. I then told the FLM, whose name escapes me, but I know had been on Jaguars for his entire ten-year career, to start the BF service. The pilot was not happy that we were not using the printed instruction manual and accused me of not taking things seriously. So, I got out the manual and read out each line to the FLM, who dutifully carried out the check with the pilot looking on.

He even made a point of pointing to a hydraulic pressure gauge and asked, "This one?" which was funny, but did not help the pilot's mood one bit.

So, a BF servicing that should have taken twenty-five minutes took nearly an hour, but at least I did learn quite a bit more about the Jaguar systems than I knew before.

Back in the Land Rover, as we parked outside the operations block, the pilot turned to me and, obviously, realising that he had made a bit of a twat of himself, tried to justify his actions by saying, "Corps, you have to realise that this mission is of the utmost importance and the service procedures are approved by Number 10 itself."

At that point, I just could not control myself and I burst out laughing. One hour later, I had been replaced on QRA and was stood to attention in front of the Squadron Wing Commander, with the upset pilot, our Senior Engineering Officer (SENGO) and WO Engineering all in attendance. The boss gave me his sternest disapproving look and demanded that I explain my actions and appalling attitude to what a very serious duty this was. So, I explained about the door slamming, the FLM having ten years' experience and that though I was the supervisor, I did not have the technical knowledge to know what he was doing.

The QRA pilot could sense that this was not going well for him and jumped in with, "What about your insubordination in the Land Rover Corporal?"

To which I said, "I'm sorry I laughed, sir. But when you said that Number 10 were involved in writing the service manuals, I just got a mental image of Margaret Thatcher turning to Geoffrey Howe and saying, 'should we check number one hydraulic gauge first or number two?'"

There were several sniggers behind me, and the boss just said, "Dismissed, Corporal."

I left as quickly as I could and was soon back in the QRA. It was a different pilot each day and I, until I returned to normal duty, didn't see the QRA pilot again, until he collared me outside the flight line hut and warned me that I was a smartarse barrack room lawyer, and he was going to make sure that I paid for embarrassing him in front of his boss. The pilot, let's call him Dick (because he was one), was true to his word and when one of my armament team failed to load one of the 30mm Aden guns properly, and it jammed due to a misaligned round in the ammo tank that should have been spotted.

He was delighted when he realised that I had signed the paperwork as the supervisor, and he made sure that the WO Engineering knew about my incompetence by telling him loudly in the crowded flight line hut. Nothing happened and one of the pilots that I was friendly with told me that the guy was a wanker and that none of the aircrew liked him, but that he was indeed gunning for me. But revenge was soon at my disposal; just a few days later, a six-ship sortie of Jaguars was about to depart but Dick's plane had an engine problem, so he needed to take the spare aircraft. As the other aircraft took off, Dick was firing up the spare Jaguar, but that turned out to have a problem too. So, it was decided that as this was an important mission, the T-Bird trainer Jaguar, which had two ejection seats, would be used.

Our team rushed out to bomb it up with 3kg practice bombs and load the guns. And as the back ejection seat was not going to have anybody sat in it, then an 'apron' had to be fitted to hold all the seat straps in place. So, I did this and then got one of the lads to quickly check it so they could sign as having fitted it and I would sign as the supervisor. The back seat of the T-Bird Jaguar is where the instructor sits and so has the master override controls when a student pilot is sat in the front seat. But, if only one aircrew is flying, then they use the front seat, which they are also responsible for; they have to make sure that the master armament override switch was set from 'Safe' to 'Armed' and normally, the armourer would do this for the pilot. But, when I saw pilot Dick running down the pan towards the aircraft, I chose not to do this. Sure enough, Dick just jumped into the front seat and ladders were removed as he shouted at the ground crew to hurry up.

Soon, he was airborne and chasing the rest of his colleagues to the bombing range. On arrival, he would find that none of his weapon systems were working so he couldn't drop any practice bombs or fire his guns. I waited in the flight line hut, where the pilots signed the aircraft in and out and waited for Dick to come in. The door burst open and in came five excited young men, talking about the mission and target that they had hit. Then, in came Dick. He walked over to his aircraft F700 paperwork folder and turned straight to the weapons loaded page and saw my signature.

He looked up and saw me and shouted, "You a-f*cking-again. All of my weapons failed and looking at this, you are responsible, Corporal."

I walked towards him and said, "All of them?"

To which, he replied with a sneer, "Yes, Corporal, all of them. What have you got to say for yourself this time?"

The room had gone quiet at his outburst, and I said fairly quietly, "You did check the Master Armament Switch in the back seat, didn't you, sir?"

As the look of realisation came over his face that this was his responsibility, and that I had probably set him up on purpose, he opened his mouth to say something, but all the other pilots just started laughing and shouting out things like *Oh dear, schoolboy error Dicky*.

He just signed the paperwork and left. I wondered if this battle with Dick the pilot would continue, but it ended there and thankfully he was posted away a few months later.

Two weeks before Christmas, we had a detachment to RAF Machrihanish, near Campbeltown, on the Mull of Kintyre in Scotland. It was as wet, windy and cold, as you would expect in that part of the world in December. To make matters worse, we were living on the base in a low-level block house that resembled a public toilet. When not at work, we went for a hike down to the headland, where a notorious helicopter crash would occur ten years later. We also visited the local watering holes in Campbeltown. One night, I was with Billy Anderson, who I should explain, had form with taking me to all sorts of historic places in Scotland, including Glencoe, and lecturing me on the evil deeds of the English. Although I was personally blameless, as I was not there or responsible in any way, I could see that he had a point and that the battles had been bloody affairs. I can only imagine what Billy was like when Braveheart came out in 1995 with Mel Gibson and I would not be surprised if there is a photo of Nicola Sturgeon on his mantlepiece. I recall that as we went into a pretty rough pub in Fort William one time, it had the proverbial sawdust on the floor.

As we went inside, he said, "By the way PJ, you dinnah wanna speak in here as they hate the f*cking English."

So, when a local gentleman asked me the time at the bar, I just showed him my watch – much to Billy's amusement. So, now, we were in a bar in Campbeltown and our armament chief, another Scot, who was sat with two older women, saw us and called us over. He introduced us as 'his boys'. It turns out that one of the ladies is from the local and all-powerful Campbell family and later, I would see exactly just how much power she wielded. I don't remember her name, so let's go with Mary. We had a few drinks with them and, being the token Englishman, I was getting a right roasting from all four of them about how the English were ruling their lives from London, and how these ladies were fully paid-up members of the Scottish National Party.

I was taking it all in good humour (as if I had a choice), when Mary's friend said something like, "PJ, do you know we hate the English because you have lots of motorways in England? It takes us four hours driving to get to Glasgow from here."

I immediately saw the look of concern on Billy's face when she confirmed to me that it would be great to have a six-lane motorway running down the Mull of Kintyre as, apparently, going shopping in Glasgow was more important than the natural beauty of the landscape. Naturally, Billy was appalled! We were invited to go back to Mary's apartment in her freezing cold open-top Porsche 911 and she just waved at the police patrol car as we hurtled down the high street at over 70mph, to my surprise, he just smiled and waved back in a sort of 'Evening, Miss Campbell' way. Mary's apartment was very plush, and everyone had plenty more drinks. We all had a good time with lots of laughter and, at the end of the evening, she ran the three of us back to the camp.

As we got out, Mary said, "I want you two boys to join me at my company Christmas Ball on Saturday," and gave me two tickets.

Our chief was not invited, which was a bit awkward! Come Saturday, Billy and I were in a taxi from the base, heading to the largest venue in Campbeltown, which, it turns out, was hosting the Campbell's fish processing plant's annual staff Christmas party. Our tickets gloriously announced this event was called the '1983 Fish Gutters Ball'.

We walked in and it was immediately as if we had severed heads; the room went quieter and we just stood by the entrance to the hall, which had set tables everywhere and a dancefloor near a stage at the front. There had been a bit of trouble between our squadron lads and the local boys, who, no doubt, would not be keen on their local girls interacting with us. Two of them started walking over towards us, full of menace, about to enquire about our presence, our parentage and to tell us how welcome we were. But, as they closed in with clenched fists by their sides, Mary came across and said hello to us and gave us both a peck on the cheek.

The two local lads stopped in their tracks and looked uncertain when Mary asked them, "Is there a problem here, boys?"

The transformation from threatening thugs to timid young boys was startling. "No, Miss Campbell," they answered meekly.

Mary smiled and said, "Good, because these young gentlemen are friends of mine and are our guests tonight."

With a sheepish, "Yes, Miss Campbell," the two lads scuttled away, and Billy and I suddenly felt invincible as we were led to a small table set for just the two of us.

The night proceeded and the place got louder and louder with the music, laughter and people generally having a good time. Billy and I made the most of the free bar. We were first to open our crackers and don the coloured paper hats before we loaded up our plates from the buffet. For a laugh, we re-enacted a Blues Brothers film scene, by tossing the gorgeous seafood bits into each other's mouths across our table. A while later, Mary came by to check on us and we said that we were just fine.

Then, she asked if we liked the look of any of the girls at the party and Billy said, "Nah, I'm just enjoying my beer and this food, which is great by the way. Thank you for inviting us."

Mary then asked me, "What about you, PJ? Do you fancy anyone here?"

Being married, I should have just followed Billy's line, but instead, I absentmindedly just looked across the floor and nodded at a pretty girl on one of the tables and said, "She is nice," and I thought no more of it and resumed having a laugh with Billy after Mary had left.

A little while later a middle-aged lady came over to me and told me that her daughter wanted to dance with me and would I do her the honour. So, I followed the lady back to the table and the pretty girl I had pointed out got up and took me to the dancefloor. We had a dance and chatted a bit, but I could sense that she was feeling really uncomfortable, so straight after the dance, I thanked her and went back to sit with Billy. About twenty minutes later, the mother was back saying her daughter would like to dance with me again, which I thought was weird as she had not seemed that bowled over about me the first time, and why was her mother asking me? This time, as I approached the table and the girl got up, I noticed a rather glum looking lad sat next to her. On the dancefloor I asked about him and was surprised to find out that it was her fiancé, and they were getting married next spring.

So, naturally, I asked, "Well, why are you dancing with me then?"

She explained that Miss Campbell had asked her to look after me and to do anything I wanted. The penny now dropped for me, and I asked about what exactly 'looking after me' meant and did it include spending the night with me. Apparently, it did, but although she thought I was an attractive guy (did I mention her dark glasses and guide dog), she would prefer to be with her boyfriend. She also made it clear that she had no choice as her entire family worked for the company, as did her boyfriend's family.

I was embarrassed that she had been put in this position and I took her by the hand and walked her back to her table and said to her fiancé, "Listen, you two go for a dance and don't worry about anything and I'm so sorry this happened."

The fiancé just nodded, but obviously felt totally emasculated and I really felt for him. I went back to sit with Billy and made the huge mistake of telling him what had just happened. As he grasped the situation, his jaw dropped open and then he looked angry. This happened just as Mary came back to our table to check on us again.

Mary was just starting to talk when Billy stood up and shouted, "You dinnae treat people like that!"

And what was left of his pint had soaked her dress from her breasts down. I don't know if he knocked his drink over jumping up to confront her or deliberately threw his drink at her; neither does Billy, but I do rather hope that he did it on purpose and that my memory is not enhancing the story for dramatic effect. But either way, Mary was now very wet and looked totally confused as she did not understand what Billy was on about, but I knew that our invincibility cloak had just been dissolved by a pint of lager and so I just grabbed Billy.

I said to Mary, "Thank you so much for inviting us, it was really nice, but we have to go now," and then I dragged Billy rapidly out of the door.

I could imagine that as soon as Mary realised that she had been humiliated in front of all of her employees that she may just unleash the thugs on us that she had restrained at the start of the evening. By now, Billy's anger had subsided and suddenly, he too realised the precarious position that we were now in. Fortunately for us, there was a taxi for hire parked across the road and we dived in and were off back to the RAF Machrihanish camp. I did not relax until we were back inside the camp gates as I thought, if Mary had the local police in her pocket, then she could probably easily influence a taxi driver to return us to the party venue, where there were way too many people skilled in handling fish knives for my liking. Billy and I didn't leave the camp for the final two days of the detachment and as we took off to fly back to Bruggen and flew over Campbeltown, I was reflecting on our strange night at the fish gutters ball and how it had all been like some plot line from the Dynasty or Dallas TV shows.

The following year, the Merc tank failed its German equivalent of the UK MOT, and it would not be viable to repair it. So, along with the VW GTi, we traded it in for a brand-new right-hand drive Volvo 360GT hatchback that we would also take back to the UK when our tour ended the next year. After the GTi, the Volvo was not the most exciting car. I even called it 'Marvin' after the paranoid android in the film 'The Hitchhikers Guide to the Galaxy', but it was a very practical car. During 1984, there was a transition of the squadron from operating Jaguars to having the new Panavia Tornado multi-role combat aircraft.

It was quite strange for several months, having two lots of squadron folk operating in the same facilities. In November 1984, the squadron was disbanded as a Jaguar squadron and there was a final handover to the new Tornado squadron guys. Other than a few key people that were being transferred to the new squadron, most of us were all posted away. It felt like the end of an era, which, with the Jaguars leaving Germany, it was. I loved living in Germany. I loved the food; the football and we had enjoyed an amazing social life. Having a bit more money had meant that we could travel around Germany, Holland and Belgium; and, knowing we were on a fixed timetable meant that we actively tried to make the most of each weekend and get out and visit places and have cultural experiences. At that time, I probably knew more about Germany than I did my own country!

This was the one time my posting dream sheet had actually worked, and I had got my first choice posting, and so we were heading to Norfolk and RAF Coltishall. So, sad as I was to be leaving Germany, Sam was delighted she was going 'home' to Norfolk and I was happy about that too. I was going to another Jaguar squadron, which was a great result.

8. The RAF's Best Camp

RAF Coltishall is in North Norfolk about ten miles from the city of Norwich. It was opened in May 1940 as one of several stations built to counter the threat of Hitler's Germany. It operated as a fighter station during World War II operating Hawker Hurricanes, with Douglas Bader as its most famous fighter pilot. After the war, it was an air defence unit with Electric Lightnings and then became a base for three squadrons of ground attack SEPECAT Jaguar GR1 aircraft. Unlike many other RAF bases, Coltishall was never developed with Hardened Aircraft Shelters (HAS) and still had the World War II dispersals dotted around the airfield with just revetment walls. So, it remained looking like it had in World War II. Its location in Norfolk, near glorious beaches, the Norfolk Broads, and yet close to Norwich to access shopping, made it a great camp to live.

I remember a team from PMC visiting to do surveys and interviews and I was interviewed towards the end of their visit. I asked them about what they had found out and they told me that the camp had one of the lowest personnel churn in the RAF, when people came here, they were happy to stay. They also thought that the camp's compact layout and lack of HAS structures, as well as its geographical location, all seemed be positive factors. In conclusion, they thought it was probably the best camp in the RAF, so naturally, in 2006, the Ministry of Defence closed it down. It is now home to a prison for male sex offenders and a huge solar panel farm!

We arrived back from Germany in November 1984, which was at the height of the miners' strike. Not having kids or pets, we had managed to clean our married quarter flat in Wickrath so well that it gleamed. I had also polished the floors with a liquid floor polish called 'Glanzer' to a mirror finish each night for about a week before the 'march-out'. The families guy inspecting it was so impressed that he wrote a letter to the estates office at RAF Coltishall and, as a result, we were given a nice, detached bungalow in Sprowston on the edge of Norwich, about ten miles from the camp. This suited us just fine as Sam wanted to get a job in the city. The removals van arrived; we had just about got the place sorted and I had set up the TV. When I switched it on, the news was all about the Bhopal Gas disaster, which was awful to watch.

At work, on 6 Squadron, I soon settled in. By then, I had several years of Jaguar experience, and Mick Gaskell was the armament chief at the time. I don't remember that much about Mick other than he was a fair boss, and, to me, he looked a bit like John Denver; so, I was always half expecting him to break out into a song about country roads. The other armourers that I can recall included Paul Griffiths, Danny McDade, Ian (Goosey) Gandon, Stu Dyson and Kev Doyle.

I really liked our new situation. I was happy on a Jaguar squadron so there would be lots of detachments to stop me getting bored. Sam got a job she enjoyed a couple of miles from our home and was only an hour away from her parents', near Kings Lynn. Just like everyone else at Coltishall, I was happy to stay for a long time.

Fig 17. Jaguar GR1 – XX729 of 6(F) Squadron (Courtesy Maurits Even)

1985 arrived and in the spring, the miner's strike finally ended. I had been going rough shooting with Tosh, my father-in-law, and his black Labrador 'Rocky' most weekends; he had bought me a shotgun as a Christmas present. As we were settled, and because I was a shift-worker on the squadron doing either days or normally early nights, it seemed the right time to get our own dog. We went to a breeder in East Norfolk and there were five small adorable black Labrador puppies for sale. They were playing with each other, rolling around and scampering about; it was impossible to tell them apart or make a choice of which one to buy. They were a Kennel Club registered litter and sired by Sandringham Sid, who seemed to have fathered most of the black labs in Norfolk!

I heard it said that a dog chooses its owner and maybe that is true, as one of the male pups saw us and left the others; he came to me, squatted and tried to pee on my foot. And if that was not enough, he took a few steps away, and did a little dump on the grass before coming back to me for a stroke and some attention. Benson had just introduced himself to me and it was the start of a relationship with an animal that I have never experienced since.

Because I was on shifts and I was at home so much, Benson really became my dog. I was training him like a gun dog, but he was also a pet, and so we went out on walks together, played football with a tennis ball and I wrestled with him all the time. I have heard people with dogs, but no kids, say that the dogs are their babies and that it's like having kids. Its only when you have actual human kids that you realise how ridiculous that statement is, but at the time, that was us. The downside to Benson as a puppy having lots of attention was that when I was working days or away with the squadron, and with Sam at work, he was left on his own and got bored and destructive. He chewed the legs on one of the dining chairs, Sam's shoes and even a credit card that had arrived in an envelope by post.

Fortunately, this phase soon passed, and he grew into a magnificent specimen, and I was offered hundreds of pounds when out walking with him to let him sire other people's Labrador bitches. But Tosh warned me that once he had a taste of that, I would never keep him in; so, sadly, for Benson, that didn't happen for him. He also became a good gun dog. Of course, the desire to track and retrieve is built into their DNA but he was a natural. I will just tell two stories here to demonstrate that.

The first was when we were on a farmer's shoot, in which you were either 'beating' and walking the fields to scare the pheasants into the air and away from you towards the other half of the party, who were in a straight line at the end of the field of sugar, beet ready to shoot the approaching birds, which by then would be over a hundred feet high.

As I said earlier, this was not like the commercial shoots where thousands of pheasants are bred on estates and a massive line of beaters force hundreds of birds over people that have paid thousands of pounds to shoot lots of birds. This was a bunch of farmers and country folk that paid in some money to buy about five hundred pheasant chicks that would be grown in open pens and then allowed to roam the local area. At the end of the shoot, we would all have walked miles and, if we had thirty birds to show for our efforts, then that was a good day.

The other major difference was that the birds shot would all be taken home by the farmers, prepared, cooked and eaten, unlike some commercial shoots that have been known to bury the unwanted bird carcasses. I was a standing gun on this occasion and the beaters were approaching; suddenly, about ten birds took to the air and were climbing towards us. I fired my first barrel and shot a bird cleanly, and it was dead before it thumped into the ground.

With my second shot, I hit another bird and it fell, but then became a 'runner'. I hated when that happened as the poor creature may now be suffering. Of course, if I didn't want to hurt animals, then I should stop shooting all together and as I grew older, I found myself more interested in watching Benson working than the actual shooting. The drive ended and Benson dutifully retrieved the first bird and brought it back to me. Then, he just ran off into the sugar beet field and ignored me calling him back. I was shouting for him to return and getting angry that he was being so disobedient.

Of course, the Norfolk farmer lads loved this, and one of them said, "Can't you control your dog, boy?"

By now, Benson had a new rude name and I was screaming at him to come back, much to the amusement of the farmers. Then, about two hundred yards away in the middle of the field, a pheasant suddenly popped up and dropped down again, then it popped up again, but this time, a certain black Labrador caught it mid-air in his mouth and then disappeared below the sugar beet; it was like the opening scenes of the 1960s children's program, Stingray. About two minutes later, Benson came out from under the sugar beet at the edge of the field and dropped the now dead bird gently at my feet, sat down panting and looked up at me.

To which, one of the farmers said, "So, boy, is your dog no longer a f*cking twat that you are going to kill when you get a hold of him?"

And everyone laughed, including me, as I patted Benson and told him he was a good dog. I never doubted him again as when he thought there was a bird there, then there usually was!

The second occasion worthy of telling was when just Tosh and I were duck shooting one evening over in the Cambridge Fens. We had stealthily approached a dyke and walked on a few ducks. Startled, they took off and we managed to shoot a couple of them. We were now just sitting about a hundred yards apart, along another dyke, in the peaceful twilight just after a glorious sunset, just in case any late birds flew in to settle in our dyke.

We were close to leaving when Benson suddenly tilted his head to the right and then looked along the dyke and then up, then at me before looking along the dyke again. I thought, *Is he trying to tell me a bird is coming?* Just in case, I lifted my gun and took general aim in that direction. Then, I heard the faint beat of wings and looked up and saw a solitary duck, way up high, flying towards me. I knew it was too high for my left barrel so I went with the chocked down right barrel that would fire the cartridge lead shot higher. I could only just see the bird in the gloom as I aimed way ahead of it to give the bird a massive lead-time to hopefully allow it to fly into the shot and I fired. The bird was directly hit and dropped like a stone into the dyke with a splash. Benson dived in after it and retrieved it and we started walking back to the car.

I told Tosh what I thought Benson had done and he replied, "Good job, you got it then. And by the way, that was one of the best shots I have ever seen in my life. Well done!"

I felt so proud. From Tosh, that was high praise indeed. But I knew that without Benson tipping me off, I would never have had enough time to prepare for that shot of a lifetime.

The summer of 1985 saw the USA for Africa and Live Aid Concerts with Bob Geldof screaming at us all to donate money as he sorted poverty in Africa out. I also got another motorcycle as I really missed by old BMW. Tosh had managed to get me a second-hand Honda 250cc for about £200 and I used it to commute to work and back, but it really lacked power and, on a visit to a motorcycle dealership in Norwich to get a new jacket, I saw my dream machine in the window—it was a BMW RS1,000 tourer and I just fell in love and wanted it so badly. I asked about it and apparently, it belonged to a veterinary surgeon who had just traded it in. I sat on it and just couldn't help myself when they offered me a low interest finance deal and a good trade in on the Honda. They also included the MOT and a year's comprehensive insurance as part of the package, so I just had to find about £250 in cash to seal the deal.

Fig 18. The bike I just could not resist and bought within 20 minutes of seeing it!

Anyway, an hour or so later, I was riding my new BMW dream machine home. This had been the ultimate impulse buy and I had not even discussed it with Sam; it was not as if we were loaded with money and so I knew she would quite rightly be pissed at me. In the end, I decided the best way to tell her was by bravely riding up to her workplace, getting her attention and that of all her work colleagues, by banging on the window. She looked and I smiled lamely and pointed at what I was sat on. Sam did not like drama or making a scene and so she just smiled and nodded and then came outside to see me and my new acquisition. I explained what had happened and apologised and she stomped off, clearly not happy. By the time she got home after work, she had calmed down and the price for my misdemeanour would be a holiday in Cyprus. As I would also be going on this holiday, I thought I had got off very lightly indeed.

The role of the Jaguar squadrons at Coltishall was very similar to that of the Harriers in Germany, which was to attack Soviet tanks and slow any invasion down until the Americans and rest of NATO came to take over. So, in reality, just the Americans then. Six Squadron's war role was in Denmark at a Danish Air Force base called Tirstrup near Aarhus, and we went there once or twice a year for war game exercises. The first time I did it was when we flew in during a TACEVAL exercise in a Hercules transport plane, and because we were practicing war scenarios, the pilot had to do a tactical landing.

The pilot did warn us, but the words "I'm going to do a tactical landing, so hold on!" did not really prepare us for the gut-wrenching experience—the pilot basically turned the plane on its side so that it vertically fell five hundred feet out of the sky in seconds, then straightened it out before thumping on to the runway and screeching to a halt as quickly as possible.

We leapt out with our guns and took up tactical defensive positions around the plane until it turned around and roared back down the runway and off up into the sky. I can remember lying there, trying not to vomit, and wondered if my guts were still in my body or were now flying back to the UK in the Herc – welcome to Denmark!

Fortunately, this mode of violent transportation to Denmark was not our usual way. We normally travelled in lorries, carrying our equipment and buses by ferry from Harwich to Esbjerg on what we called the 'party boat'. Our food was included in the ticket, and we had beds in a cabin too, so it was just a great overnight cruise to us. There was a small casino and a tiny disco area that was open until 2:00am, but my favourite memories are from our times in the piano bar. There was usually a crooner guy, but occasionally, a small band—and whichever it was—were normally not always great to listen to and were probably real cruise ship rejects. As such, they were a target for our good-natured banter that today would probably be regarded as abuse and perhaps should have been then. They normally took it with good grace and even handed us some back, which made it fun.

But on one occasion, the band's lead singer lost the plot completely and stopped playing and said, "Guys, this job is not easy and if you lot think you can do better, then come up here and have a go."

I guess that his 'put up or shut up' line would have worked most of the time, but he was unlucky as we had several guys that could play instruments and one was a pretty good singer, so they decided to accept the challenge. A look of surprise on the band's lead singers face was replaced by a smirk when the guys picked up the instruments and looked at each other, asking what they should play. They explained to the audience that they had never actually all played together before, but they had agreed on a song and that song was 'Hotel California' by the Eagles and off they went. It was a revelation, not as good as the Eagles obviously, or even a cover band, but it was good enough, and we cheered them on; the other passengers joined in. At the end, everyone cheered, clapped and whistled and our guys were getting a taste for it and asked for requests, but that was too much for the ship's band leader.

He shooed them off the stage with a reluctant, "Thank you, guys. Very good, but we will take it from here," which got resounding boos from everyone in the audience. I bet he never used that challenge with hecklers again.

Years later on a ferry trip from Europe to the UK for a Harrier detachment, Haggy, one of the lineys, made himself a star. We were wondering why he was sat by the lone piano man and chatting with him between songs for about twenty minutes.

Then, the piano guy said, "Ladies and gentlemen, we have a young man here from the Royal Air Force and he wants to play and sing you a beautiful song. It's my pleasure to introduce you to Haggy. So, please give it up for Haggy!"

The other passengers clapped politely, but we were watching in silence as we didn't know that Haggy could play the piano in addition to his other instruments.

Haggy assumed the seated position and said, "Thank you, Ron, for this opportunity. Ladies and gentlemen, this is one of my favourites and I hope it's one of yours."

Then, he simply clumped his hands up and down on the keys and the piano made a dreadful noise, the passengers looked stunned, the ship's pianist looked horrified, and all the squadron guys just fell about laughing. So, Haggy didn't play the piano after all!

My other party boat memory on another trip was when the bar closed and everyone was heading off to their cabins, one of the guys got into the information booth and operated the tannoy system and gave a 2:00am announcement throughout the ship, *Ladies and gentlemen, we have a special offer on this crossing—half price wet fish on the car deck, wet fish on the car deck, first come, first served*!

The ship would dock in Denmark at around 7:00am and then a bunch of very hungover and tired RAF lads would drive the three-hour drive to Tirstrup near Arhus, almost certainly over the legal alcohol drinking limit. It's funny though, when going home again, the sailing back to Harwich was always a much quieter affair. A different mood tone and much less drinking, probably because we were all tired and resetting to go back to our families. So, the party boat effect was only a one-way affair. On the weekends, we would go off drinking locally in

Ebeltoft or Arhus and even Randers, but before we left for any night out, we made a point of drinking a few extra cans of beer in our camp as the cost of a pint in these Danish towns was extortionate, even with our extra LOA money. One night, I was in an Arhus nightclub with Morgan 'Geordie' Ward, who was an avid Sunderland FC supporter and a doppelganger of the A-Ha lead singer Morten Harket, which meant he had girls very interested in him—something he didn't mind at all!

On this occasion, he was in the middle of the dance floor in a fight and shouted for my help. I piled in and got a punch in the throat for my troubles, and then there was a bit of a stand-off. During this pause in the action, I asked Geordie what had started it. When he said that the "Bloke over there didn't like me staring at his girlfriend," I realised that we were not on the side of the angels here. So, I waved my arms to calm things down and offered the lads we were fighting a beer and said that it was all just a misunderstanding. We then got a super expensive round of beers, which Morgan had to pay for.

We had once been in Gibraltar on a 'show of force' mission to the Spaniards after the Falklands invasion gave them ideas about invading Gibraltar. After a long bus ride, we had ended up in Torremolinos and managed to convince everyone that we were professional football players that played for the German team Borussia Monchengladbach. This only became an issue when a sceptical English street ticket seller girl challenged us to talk with a bunch of German fans. So, I did one of the team's chants and they responded with huge cheers and joined it; by chance, they were from Monchengladbach, or at least knew the team's chant!

The English girl responded by saying, "Bloody hell! You really are footballers," and gave us free entry tickets and drink vouchers to one of the nightclubs. We soon found out that Spanish girls and those on holiday from all over Europe were fans of Morten Harket too!

In 1986, Haley's comet made one of its 76-year appearances and sadly, seven astronauts lost their lives in the Challenger Space Shuttle disaster due to an O-ring seal failure. After my annual assessment, I had been called in to see the JENGO and was told that I was being considered for a place at Shrivenham, which is a military school at which I would be posted to and study an engineering degree, and then become an engineering officer. This really appealed to me but first, I would have to get three more O-levels and pass the officer selection process at RAF Biggin Hill. So, I visited the education centre on camp and signed up for three O-levels, taking Physics, Economic & Public Affairs and General Paper through evening classes.

A few months later, I managed to pass all of them with a 'C' grade, so I now had five O-levels. I then went off down to Biggin Hill for the officer and aircrew selection process and got through the three days of tests, interviews and physical team tests. I was then told that I had not been successful for an engineering officer, but as I had demonstrated good logistical skill traits, I would be considered for a commission as a supply officer in a couple of years' time. But for that too, I needed to have a better grasp on politics and current affairs.

Now, you may have noticed that I have not exactly been kind regarding the investigative qualities of members of the RAF Police. Having dealt with quite a few members of that branch over the years, I had my suspicions, but it was at Coltishall where they were confirmed. There was a story about a guy (I think also an armourer) on a different camp, I think told to me by Stu Dyson that was into his motorcycles, but he had needed a new engine for his. He was a member of the camp motor club and when he got a new engine, he fitted it and rode out of the club. As it happened, there was another motorcycle being stored temporarily in the motor club at this time and its engine was stolen around the same time as this guy fixed his bike. The RAF plods investigated, and within just a few minutes, they had cracked the case—the armourer had obviously stolen the engine. Naturally, this resulted in a disciplinary charge for the guy with the most likely outcome being a prison sentence at the military prison in Colchester.

At the first charge hearing, the JENGO as his flight commander deemed that the case was so serious it needed to be heard by a higher ranked officer, who also had greater punishment level powers. At the hearing with the SENGO squadron leader, the armourer refused to plead on the charges and utilised his right not to accept the punishment of that officer and instead demanded the right for a court martial, which is like going to the magistrate's court in civilian life. This was unusual as opting for a court martial as it would mean a far stiffer punishment if found guilty. Apparently, the RAF Police were getting nervous about the situation as a court martial involved great cost; it took time and involved military legal teams as well as the camp's senior officers. Eventually, the officer commanding the RAF Police and the squadron senior engineering officer (SENGO) called in the armourer for a meeting, wanting to know why he was so relaxed about having a court martial. Apparently, this armourer quite liked being a bit of a maverick and as he was not a career guy, he was always happy to be awkward if he fancied it. The SENGO knew that the guy would not risk a prison sentence and so thought that the RAF Police had got something wrong. The armourer stated that the motorbike that had its engine stolen was a Kawasaki and had a 250cc, two cylinder, two stroke engine and that the RAF police officer agreed with that statement of fact.

So, then, the accused armourer asked, "So, as I own a Honda motorcycle with a four cylinder, four stroke 400cc engine, what use would I have for a smaller engine, of the wrong type, that would not fit or work in my bike anyway?"

The story goes SENGO looked at the RAF police officer and said something to the effect of, "You have got to be kidding me? Your investigation did not reveal that the bikes were not the same? Thank God this case never got to court martial, we would have been a laughingstock."

I would love to say now that the RAF police officer still did not get what the problem was, but apparently, he did, and he had to apologise to the armourer and the SENGO for wasting their time. I would hate to have been the RAF plod that did the investigation and brought the charges.

During that summer's annual APC to Decimomannu in Sardinia, I had the chance to fly in our T-Bird trainer on one of the bombing sorties. I had my thorough pre-flight medical, which delved deeply into my health status by making sure they had spelled my name correctly and asking if I had a cold! Then I was fitted out in a flying suit, Anti-G suit and a helmet and I was ready to go. I attended the flight briefing about the mission—it was just a simple flight to the range, drop some 3kg practice bombs and fly back. Just before I got into the plane, the pilot came over to me to brief me on what to expect and ask about my flying experience. He was a big bloke from near Barnsley and had a very strong Yorkshire accent. He also made a big deal about how, unlike a friend of his, he was not going to die if there was a problem and we had to eject, he would give the order, *eject, eject, eject*, and then he would pull his ejection seat handle and eject without waiting for me to go first. He pointed out that at 250 feet above the ground, at 500 knots, it was under two seconds before impact for a crashing aircraft. I told him I was on message and would eject immediately if I heard the command.

I loved the flight and feeling the G-forces in my flying suit as we climbed away after each bombing run. On the way back, we climbed up a couple of thousand feet. He let me take control and we did a few basic manoeuvres, such as steeped banked turns and finally a staggered roll with three pauses after each ninety degrees of the roll, the way the Red Arrows did. It was great, but as we flew back, suddenly the cockpit was full of alarm noise and flashing warning lights. I had both hands on the ejection seat handle; taking a deep breath, I had straightened my back, ready to pull the seat pan ejection handle and eject.

The pilot must have sensed this as he shouted out, "It's okay, it's just the LOX has run out. Just remove your mask and we will go home low-level."

LOX stands for Liquid Oxygen supply, which was processed to become the cooled air supply that we were breathing through our masks. He made the cockpit alarms go quiet, which did lower my stress levels; but as soon as I took my mask off, the warm, oily, rubbery smell of the air in the cabin hit me and I soon felt nauseous, but I managed not to puke before we landed. I swear that when those alarms went off, if the pilot from Yorkshire had said, 'Eee bah gum' instead of 'it's okay', then I would have pulled the handle and been possibly the only armourer to have ejected from a serviceable aircraft.

In 1987, the world's population reached five billion and Sam and I were trying to add to it, but nature was not playing ball. At Bruggen, all our friends had children and we were twenty-five years old, which at that time was on the late side to start having babies. I really don't know how people in their late thirties and forties today find the energy to cope with babies!

Nothing had happened after a couple of years and so we went to see the medical people about it in Norwich. They gave us charts, a calendar and a thermometer and depending on Sam's temperature and getting close to and during the ovulation phase in her menstrual cycle, we were encouraged to have lots of sex. Naturally, I hated this enforced pleasure, but despite our efforts, no reward in terms of a pregnancy, which raised uncomfortable questions about fertility. I could not help but think about that time in Sicily, lying under that Victor aircraft, getting my nads x-rayed by the plane's navigation TACAN. So, Sam had tests with her GP in Norwich and I went to the medical centre at RAF Coltishall—a big mistake!

The RAF doctor gave me a sperm sample tube and said that I had to do the business the following morning and drop it off at the medical centre before 8:30am the following day so that it could be transported to the laboratory in Norwich for testing, and that they would have the results back within two days. I was on day shift that week and started at 7:00am the next day. We armed the eight Jaguars on the 6 Squadron flight line with practice bombs and loaded the guns with practice 30mm ammo. I remember that it was a rainy and dreary day and after I had signed the paperwork, I headed off to the loo to try and produce my sperm sample. I can tell you that loading aircrafts on a wet day first thing in the morning is probably not the most conducive condition to promote feeling amorous. I sat in the line hut toilet cubical with my oily overalls down around my knees, trying to get inspired, but after reading the graffiti on the walls pronouncing that "Jacko was a wanker," I thought to myself, *he is not the only one!* Like most men though, with enough focus, we can usually 'knock one out' if we are not too drunk, and soon I had delivered my little bottle to the medical centre in time for it to be sent off to Norwich. Two days later, I had a call to visit the medical centre to get my results, and so I shot up there, eager to know my results.

What happened next could only happen in a military environment and probably broke numerous ethical codes of the British Medical Association. I walked into the reception, which was full of post-natal women, clutching small babies, waiting for some sort of training session with a midwife. I walked up to the receptionist RAF nurse on the desk and announced who I was.

She looked at the diary and said louder than was necessary, "Oh, you're here for your sperm test results?"

In a rather embarrassed voice, I said that I was.

She smiled wickedly at me and said, "Well, actually, I have your results here and I can give them to you."

This surprised me as I was expecting to meet with a doctor to discuss the matter, but now I was expecting her to just hand over the report in a sealed envelope so that I could digest the news in private. But no, she opened the envelope, glanced at it for a few seconds and then proceeded to read out the test findings. The reception area suddenly fell silent, and I mean silent. I swear that even the crying babies stopped and were looking at me. I was mortified, and I just stood there, totally stunned and feeling monumentally exposed. She spoke loudly enough to effectively be announcing to everyone that after twenty-four hours, I had forty percent dead or dormant sperm and I think she said some were misshaped and others were swimming the wrong way. There were a few "ahh, bless him," from the mothers in the room, proudly holding their babies as living proof that their husbands did not have geographically retarded, dormant or dead spunk.

But then, after a pause to amplify my moment of humiliation she said, "But you do have almost sixty percent of healthy sperm still alive after twenty-four hours, which is a really high percentage." She looked at my no doubt confused, concerned and bewildered face and said, "Basically, honey, you have shit hot tads."

I don't think I have ever felt so relieved, and as I left the medical centre, the post-natal ladies were now saying things like, "Oh, I wish my hubby had shit hot tads, it took me years to get pregnant."

On reflection, I am sure that the nurse receptionist would not have read the results out if it had been bad news and being forces, she just saw and naturally took the opportunity to have some fun at my expense. Once again, the old forces adage of 'If you can't take a joke, then you shouldn't have joined up' comes to mind.

So, with a mixture of total relief and a newfound pride in my high-grade spunk, I told Sam rather insensitively, "Well, it's not me, I have shit hot tads!"

Which was possibly not my finest partner support moment. But fortunately, it was not her either as she was perfectly healthy, which was confirmed a few months later, when the hospital called to announce that she was indeed now pregnant. This life-changing news came through on the very day we were moving house from the married quarter in Norwich into our own, recently purchased house in a small town called Stalham in northwest Norfolk. House prices were going up like mad and on a detachment, one of the Riggers called Gary Le Page told me that his house was earning more each day than he was. So, I sold my beautiful BMW bike to pay for the deposit and with some financial help from Rosie and Tosh, we bought our first house. This is where my property buying luck showed itself to be bad; we bought at the peak before 'Black Monday' caused a stock market crash and our mortgage interest rates suddenly shot up and by 1990, were at 15.4 percent. Eventually, we would sell the house for the same as we paid for it – that should have been a warning to me not to invest in property, but I didn't learn.

In early 1988, I had taken a week's leave to have a holiday in Egypt. On my last day at work on the squadron, there was a bit of a panic as they were trying to launch six aircrafts on a sortie, but the armament load requirements had changed; they needed to get the kites airborne before an approaching storm hit Coltishall, which would delay the mission. The wind was already gusting strongly, and squally rain was hitting us as the team loaded the aircraft. We just got it completed in time and off they went and off I went on holiday. I came back and was called into the WO Engineering's office immediately. I noticed that everyone was looking at me in a strange way – what the hell was going on? I soon found out. One of the large, curved ammunition tank panels that was secured with about fifteen captive screw bolts had been closed but not secured by tightening the bolts; the panel had ripped off in flight and nearly hit the following aircraft. If it had gone into its engine intake, then the result could have been catastrophic. Understandably, the OC Engineering Officer at Coltishall wanted me to face a technical charge as I was the supervisor.

This was entirely my fault as I had devised a loading method for my team that did not follow the Air Publication (AP) procedures in the right sequence. Instead of doing one-gun loading operation and one-bombing load procedure as a team in individual operations, I had split up the various activities so that one person did the same job on all six aircrafts and then I would go around checking all the work before signing it off. This technique saved us about twenty-five minutes each morning and we had never had a problem. But that day, with the storm coming in, we had all been rushing about, including me, and so one guy forgot to secure the ammo panel and I didn't notice it because I had asked another JT that I trusted to check everything so that I could run in and do the paperwork.

The WO Engineering had already heard all of this from the load team and our SENGO wanted to charge the JT instead of me, but I knew that I should have checked it myself and I was also the one that signed to say it had been done. So, I just went all 'mea culpa' and refused to let the JT take the blame. Two days later, I was called back into the WO Engineering's office and I expected it to be about the technical charge. The WO told me that our SENGO had met with OC Engineering, who accepted that the pressure of getting the mission airborne before the storm hit were mitigating circumstances; but apparently, he was most impressed that I took full responsibility and didn't throw one of the guys under the bus to save my own skin. As a result, all charges against myself and the JT were dropped and that there would be no mention of the incident on my personal file, which would have been damaging for my career. I told the WO how grateful I was to him and the SENGO, and that I would also thank him personally.

The WO said, "Yes, go and see him right now as he has another matter to discuss with you."

So, off I went to the SENGO's office, totally relieved but wondering if I was in trouble for something else. In his office, we spoke briefly about the gun panel incident and then he shocked me by telling me that I was being promoted to sergeant and posted, but only to the armoury at RAF Coltishall. I just could not believe how my day had just changed from expecting a career ending punishment to getting a promotion!

It was eleven years since I had joined up and I was now a sergeant at aged twenty-eight, which meant that I could 'sign on' to serve until aged forty-seven; it gave me another nineteen years of job security, which, at the height of the recession at the time, felt very comforting. My escape from punishment and subsequent promotion was cheered by most of the squadron but not one of the fairy sergeants—what is it with those guys? He said loudly in the packed T-bar that the armament world must be really scraping the barrel to promote me. So, I asked him how old he was when he got his third stripe as a sergeant, and he told me he was thirty-two and what of it. I told him that I just wanted to calculate how much more money I, as a barrel scraper, would have than he had now in four years' time when I was thirty-two. This was thanks to my promotion being way sooner than his, and of course, the fact that armourers as Trade Group One get paid more than fairies anyway. This caused an outcry of laughter from everyone in the T-bar and I just walked out with one of the armourers who said, "Good one, PJ."

In mid-February, Sam was admitted to the hospital as the medical team dealing with her pregnancy were concerned about her health. She had developed swollen feet and that may have indicated preeclampsia. The due date came and went and finally, after many hours in labour, a very healthy boy, Stephen George James, arrived in the world, born 4 March 1988, oblivious to the sleet and cold outside at the Norfolk and Norwich hospital. A couple of days later, we went home, and I took the baby up to the new nursery that Tosh and I had prepared a few weeks earlier. I can distinctly remember holding this new life in my arms and suddenly feeling a huge weight of responsibility pressed down on me. We had bought this little human into the world and I'm sure, as every parent feels when that moment arrives, it is a feeling of being daunting and worrying if you can step-up, which of course, despite there being no manuals, most parents do.

The birth meant that I could have applied to defer my promotion courses, but we needed the money. After setting up the new house and with a baby coming, we had maxed out our two credit cards and I was only able to chip away at the debt to reduce the charges in small payments. So, just twelve days after Stephen was born, I was driving to RAF Hereford for my General Service Training 2 course that would be almost immediately followed by a Trade Management Training course up at RAF Scampton. Sam's parents helped enormously of course, and my stepmother Sylvia even came up by coach via London from Torquay to stay for a few days.

But I must give credit to Sam for managing for so many weeks on her own with a new baby. To make matters worse, as we couldn't afford disposable ones, we were using cotton flannel nappies and there was usually about ten of them on the washing line every day. We had agreed that it just was not practical for Sam to stay at work with a twenty-mile commute to Norwich every day and so she became a full-time stay at home mum.

In July, just four months after Stephen was born, we were both shocked to find out that Sam was expecting again. Of course, this was unplanned, and we should have been more careful and taken precautions, but we were naive and didn't realise that once a woman has given birth, then her body was far more ready to go again than the first time. So, now, we were preparing for another baby that would arrive when Stephen was just thirteen months old.

At work, I had moved from 6 Squadron to the Weapons Engineering Flight and oversaw the aircraft pylon servicing bay, and after just a couple of weeks, I hated it. I was so bored and felt like I was working in a factory. It was just not work that I enjoyed, I liked the hustle and bustle of first-line squadron life. The guys in the armoury were great and the place had a great atmosphere; I got to know more of the guys from the bomb dump too. The guys that I can remember. including Terry Reeves, Rick Whan, Shane Wilkinson, Marty Thorne, Mark Rudling, Andy Joslin, Ozzy Langly, Al Luckham, Chris Gregory, Tim Grigg, Phil Webb and Mal Jones as well as Sue Austin, who ran the control centre of the armoury. There were others of course, but my memory fails me.

Fortunately, I had an escape route from the Pylon Bay as the late chief tech, Pete Gilbert, who ran the Weapons Training Cell (WTC) wanted a Jaguar-experienced SNCO trainer and he offered me the role. He told me that WO Dick Bentley had already sanctioned the move. I jumped at the chance as I really liked Pete and the other guys in the WTC, but I didn't realise until later that it was also an Explosive Ordinance Disposal (EOD) training role too, and so I would have to be trained. So, that July, I was away yet again; this time, for about six weeks, on an Explosive Ordinance 8B1 course at the Defence Explosive Ordinance Disposal School in Chatham and it turned out to be the hardest thing I had ever done.

There were about sixteen of us on the course from all three-armed services. The classroom theory element alone, to learn about explosives and their characteristics, was immense and we would all be studying our copious notes late into the evenings in order to pass the regular test. Each Friday afternoon, at 3:00pm, we would be sat in class and a list of names would be read out; these names were of the people that would continue the training on the Monday. If your name was not read out, then you had a quick debriefing before being considered 'Return to Unit' (or RTU), so basically you had failed the course.

I had never seen this happen before, but it made sense as this was a dangerous job and it needed only competent people doing it. That first week, we lost three trainees, but after that, it was just one person that left each week. The practical lessons started too, and we learned new skills and how to operate under pressure. It was summer and about twenty-five degrees and for one test, I had my full DP combats on, an NBC suite complete with mask, and I had to defuse a bomb in a CS gas chamber after walking about a hundred yards carrying all my kit. I was soaked in sweat, which was also stinging my eyes that I could not rub, and I thought I was going to pass out with heat exhaustion. I also suffered a bit with claustrophobia so being confined I could feel that irrational fear starting to wash over me. So, I paused, got my breathing steadied, told myself to 'get my shit together', and then got on with the job.

There was one example of just how good German manufacturing was in World War II—a German bomb had been found inside a gasometer in London and due to its location, had to be trepanned with a disc cut into the top and then have its explosive material contents melted out with hot steam rather than just be destroyed with a controlled explosion. As a result, the long-delay fuse device remained intact and was disarmed and was now at the EOD training school. The instructor showed it to us during a class and explained that it was a twelve-volt system that, for some reason, had failed to function forty-seven years earlier. He then had the idea of testing the battery with a voltmeter and we were all astonished to find out that the fuse still had nearly eight volts left in the tiny battery!

At the end of the course, there were just seven of us left. They gave us two certificates, one for the training course and the other about being a member of the Explosive Disposaleers fraternity—that certificate is still on display in my home and is the only certificate that I have ever displayed because I felt it was so hard to earn.

A few months later, I was back at Chatham to do the Transition to War – IED 8B19 course, but unlike the first time, this was both fascinating and fun. We were taken to houses that had been adapted to demonstrate every terrorist weapon and explosive storage trick that had been discovered by the armed forces in the UK, especially Ireland, during the troubles and abroad. Some of the techniques used were both very clever and often devious.

We were split into two teams to set up booby traps in a house with sound units, which are like electronic firework bangers, and then our team had to 'clear' the other team's house of their cunning booby traps. We got through about four of the traps before being blown up and I think our house got the other team at about the same time. But for me, the highlight was our final test. I was racing across the River Medway towards Chatham Docks in a Rib, which was an inflatable speedboat. We boarded a cargo ship and I had to defuse a tripwire attached to a Thunder flash, then descend some smoke-filled steel stairs in my NBC kit with gas mask, and defuse a Soviet bomb that had supposedly entered the ship but not exploded yet as it had a time delayed fuse. The smoke was so dense that I had to find the bomb by feeling around and by touch, as the torch just illuminated a wall of grey smoke. When I found it, I was able to immobilise the piston firing mechanism on the fuse to make the bomb still armed, but now relatively safe. I came out of the stairwell and then back on to the Rib boat to race back to the shore. It was all very exciting and I felt like James Bond. In terms of the training scenario, I was wondering how a bomb dropped from a Soviet bomber could also set up a tripwire trap, but hey ho!

onourable isposaleer ertificate

DEFENCE EXPLOSIVE ORDNANCE
DISPOSAL SCHOOL

Hereby Declares That

SGT P.E. JAMES

is admitted to the Honourable Company of Disposaleers.
He is to be welcomed at all gatherings and share in the goodwill and good cheer
of all Disposaleers. He is to be accredited with all honours
and Privileges due to all Honourable Disposaleers.

This certificate is valid for the duration of the life span
of the aforementioned Honourable Disposaleer.

Signed, Officer Commanding
On the 29 day of the 7 month of the year 88

Fig 19. The only certificate on display in my home

I really enjoyed my job in the Weapon Training Cell showing armourers recently posted to the squadrons how to remove and fit ejection seats, Operational Turn Rounds (OTR) and classroom theory training. However, it was my newly acquired skills regarding EOD that was giving me the biggest buzz. There were regular trips across to Holbeach Range on The Wash in Lincolnshire to dispose of ammunition and explosives, such as IR flares and life expired ammunition but all small stuff. It usually meant having a fire in a cage on the range and just blowing up a few rounds of 30mm ammo and the odd 28lb practice bomb. I also had a trip to Shoeburyness in Essex to blow up a 1,000lb bomb, which was exciting. In the autumn, I had to prepare the airfield for a simulated attack by low-level attacking aircraft dropping bombs along the runway, and so I used steel containers with a diesel and petrol mix along with some used engine oil to create smoke. I had no time to practice this to see how they looked when set off, and I didn't realise that there was an approved procedure for doing this.

On the big day, it was raining quite hard and I was worried that the rain may dilute the mixture. So, I added a bit more fuel to each pot. The weather cleared and the attack happened. I had six guys with shrike exploders, which fire a high energy 400v pulse connected by firing cables to the detonators in the pots situated along the runway. It has instructions to fire the shrike just as the attacking Jaguar had passed them to simulate an explosion from the bomb dropped from the aircraft.

The first pair of Jaguars attacked, and as they passed, there were two 'explosions' in quick succession. Perhaps, I should have practiced before the big day as I was shocked at the result when a huge flash erupted with black smoke, and shot up about eighty feet into the sky. The first aircraft was clear, but the pilot in the second aircraft had to suddenly bank to the right to avoid the fireball coming up at him. We had similar impressive results with the second and third attack, but I did notice that the aircraft approached from a fair bit higher. It looked spectacular though and that night, it featured on the local TV news. The next day, I had a call from someone EOD at Strike Command HQ asking me if I had followed the correct procedure. I lied and said that I had, but that I had added a bit of extra fuel and oil due to the rainy conditions and that perhaps I should not have done that.

The guy on the telephone said, "Mmmm, we will have to review the procedure as the flames are only supposed to shoot up about twenty feet."

I found a copy of the procedure and it stated a maximum of a litre of mixed fuel should be used, I had used about three times too much—oops! I had a similar overcompensation experience when I was ordered to set up the 5th November bonfire at St Faiths that the wife of the Mayor of Norwich would be igniting using a shrike. I got there and it was raining hard, and all the bonfire wood was soaking wet, the locals had built the fire and told me it had been soaked in diesel hours before.

We pulled a few bits out and tried to light them and even with dry tinder, they were just too wet to burn. Luckily, I had brought two destructor incendiary blocks, which contained thermite and which we used when lighting a fire pit to burn explosives such as flares, and so I put a couple inside the base of the fire and wired them up to the shrike. In the evening, the fire lighting ceremony started, and the mayor's wife made a short speech to welcome everyone. I showed her which button to press on the shrike. It seemed that the rain had not soaked things quite as much as I thought and that perhaps two destructor incendiaries were a bit excessive. She pressed the button on the shrike and there was a huge whoomph sound, and I am sure that the entire bonfire lifted a foot off the ground. I thought I saw people's feet on the other side.

"Oh, my word, how impressive," she said as the bonfire now had flames licking the sky twenty feet above it and everyone was stumbling backwards to get away from the searing heat.

I just nodded as if everything was exactly as planned and watched the massive pile of wood get consumed in about thirty minutes. I told myself that I must learn to be more measured in these things!

Becoming a SNCO also meant that I had additional duties. Namely, I was made SNCO in charge of the accommodation block for the single armourers. Every month or so, there would be a 'bull night', but not to the level of the ones at training camps, for these the guys just had to have tidy rooms and clean sheets on their beds ready for inspection. The station commander would choose a block to inspect and one day it was ours.

As he approached and I saluted, he said, "Ah, Sergeant. Today, to make this more interesting, I want you to conduct a survey as we do the inspection to find out who is more popular, Sam Fox or Linda Lucardi."

These ladies were, at the time, the most popular topless pin-ups from The Sun newspaper and were featured in most of the single lad's rooms. By the end of the inspection, it was clear that Linda had won the popularity contest, which was a personal disappointment for me. When Sam Fox revealed that she was a lesbian a few years later, I did wonder if the guys that fancied her were disappointed or perhaps a bit more intrigued.

The year ended with a tragedy when on 21st December, the Lockerbie bomb aircraft crash happened and killed all 259 people on Pan Am Flight 103 and eleven residents of the village of Lockerbie in Scotland. Many service people were involved over Christmas and had to deal with what must have been horrific scenes at the crash site and are probably still being affected by it today.

1989 arrived and George H Bush became US President. In January, I was doing a Ground Instructional Techniques course at RAF Newton. Most people found this course hard, but not me. After the EOD course, this was a walk in the park. Also, I found out that I was quite comfortable being out front of the class doing the teaching and I was good at preparing interesting material. I was in the sergeant's mess bar early most evenings and the barman did not believe I was on the GIT course, as it was known.

At the end of the two weeks, I passed with an A1 and was called into the course leader's office, congratulated on my performance, and told to expect a posting to the training school at RAF Cosford within a couple of months. I was horrified and said that I would hate being a full-time instructor. I was told that all A1 students automatically got posted to a training school, so I pleaded with him to downgrade my results so that I would not get posted. In the end, shaking his head, he agreed, and I was now a B1, and I never did get posted to a training school, which would have killed my soul as that sort of routine work is just not me.

On reflection, 1989 turned out to be a world-changing defining year, even if not appreciated at the time as on 12th March, Tim Berners-Lee changed the world by inventing the World Wide Web and in April, the Tiananmen Square protests started and lasted two months, which would shape how China was to be governed in the future. Even more significantly, in November, the fall of the Berlin Wall and collapse of the Soviet Union took place.

For the James family, the main event of the year happened on 27th April, when Christopher Liam James was born, just thirteen months after Stephen. Unlike the protracted affair with Stephen's birth, this was over in a matter of minutes. It was so fast that there was no time to get Sam to the hospital, and luckily a neighbour, who was a midwife, came around to help. Sam was sat on the loo when it started and refused to move. I had visions of the new baby entering the world with a splashdown in the bathroom toilet. The midwife got Sam to go to the bed, and just twenty minutes later, it was all over.

Now, we effectively had a new baby and a toddler to look after, as Stephen was an early walker and so, a real handful. Soon, we just treated them like twins and that continued until they grew up—this was not really that fair on either of the boys, but then it was just about survival and getting through each day. Despite lots of family support, we were exhausted. Nature designed humans to have babies when young for a reason!

We managed, just like most couples do, but our lives had been turned upside down and Sam had already given up work to focus on the boys. The cost of another baby on the back of buying the house in Stalham and setting it up was crippling. My book system was in chaos with both credit cards maxed out. To help with our dire money situation, she also took on a part-time estate agents' job in the local town at weekends and I would look after both boys. This also gave her a short break from childcare duties. I was very glad that the roles were not reversed! I can remember at the start of one month, when I did the accounts, we were going to have to live on £45 that month. Rosie and Tosh helped out with food and money when they could, but we could not sustain this, so we agreed on two things—one, I would apply for a posting to Germany where we would get LOA money and we could also rent out our house to pay towards the ever-increasing monthly mortgage payments; and two, despite hating the idea of the role, I would try and get commissioned as a supply officer as that would mean a pay rise. My applications for both went in the following week.

In January 1990, I was told that I had been selected as one of the four RAF Coltishall personnel to be part of an exchange with the brand-new Royal Navy warship HMS Norfolk, which was a Type 22 Frigate. Four of the ship's crew would be employed in various roles across RAF Coltishall for a week, whilst we were at sea doing handover trials.

A week later, and I'm on the train with three guys heading down to Poole in Dorset to meet the ship. We were due to board a small vessel at the harbour that would ferry us out to the ship at 7:00am, so it was probably not the best idea to get a bit wasted in several pubs the night before; but, hey, ho! These things happen. I can remember standing on the harbour wall, not feeling great as the wind and sleety rain bit into our faces. The 'Liberty Boat' rocked up dead on time, and soon, we were heading out to the ship. As it was windy, I had taken my beret off in case it got blown over the side. We climbed up a net and into a hatch in the hold and followed one of the seamen welcoming party up to the bridge. I should have put my beret on then but just didn't think about it. The captain of HMS Norfolk was outside the bridge, and we were taken to him. I did stand to attention but now realised that I could not salute without wearing headdress, and so I was faffing about, putting my beret back on.

The captain was clearly not impressed and said, "Don't you salute Royal Navy Officers in the RAF?"

My hungover brain said that we didn't have any navy officers in the RAF, but fortunately, this time my mouth said, "Apologies, sir, and thank you very much for inviting us aboard. We are all so excited to be here," and I gave him a really smart salute.

That seemed to pacify him, and instructions were given to take me to the chief petty officer's mess, like our sergeant's mess, and the others to the seaman's sleeping quarters below decks. Later, we were given a tour of the impressive ship and then we set off for Land's End before heading up towards the Isle of Man in the Irish Sea. Sailing around Land's End that night, through a winter storm with thirty-foot waves, made even the large warship twist and turn and rise and fall. I was glad to be in my bunk bed and not trying to move around the ship.

The next day, as we sailed up the Welsh coast, the ship carried out an exercise and we four RAF lads with two navy guys were lowered in a rib down the side of the ship into the considerable swelling waves. The navy lads released the rib inflatable boat from the hoist with perfect timing, and soon we were bouncing across the sea and circumnavigating the ship, which looked enormous from down in the small rib.

Despite it being freezing cold, it was like being on a mad rollercoaster. After about twenty minutes, we were ready to get hoisted back up. Once again, the navy boys connected us to the hoist with precision timing, just as the swell under the rib lifted us enough to put a bit of slack in the hoist cable, and then we were swinging our way back up to an entrance in the ship's side. There were two more days of exercises and tests of the ship's systems and then it was time to head to port in Glasgow so that the ship's builders there could deal with all the 'snags' that the crew had identified after six weeks of testing. This also meant that the crew got some downtime. And, after a solid six weeks at sea, training, they were ready for it. As we sailed up the river Clyde, you could sense the excitement building. The married guys headed home but the single guys were ready to 'party' and boy did they know how to party hard.

Having saved up their money over the past few weeks, they now splashed the cash, starting with hiring taxis to be their chauffeurs for the entire evening, with them waiting outside each venue whilst we drank inside, before heading to the next pub.

As we left the dock gates and piled into the waiting taxis, the driver asked, "Where to lads?"

To which, the navy lad in the front simply pointed to a pub about one hundred yards up the road and said, "There, for starters."

Several hours and many pubs later, we were in the middle of Paisley. The taxi was finally paid off and we piled into a nightclub. It was a great night with much dancing and it was obvious that the navy lads, whilst not dismissing the interested local girls, were just quite content with having a good time with their own group of close mates; this also reduced the chances of any problems with the local lads. The night got late and the dancing, more extravagant, and I noticed a navy guy that had taken his trousers and underpants off and now had a copy of the Sun newspaper clenched in his buttocks that had been ignited. He was burning merrily away as he strutted about, performing a very credible rendition of that forces' favourite "The dance of the flaming arsehole." He was enthusiastically encouraged by the thirty HMS Norfolk crew around the dancefloor and clearly appreciated by the locals too, who, being Scottish, obviously approved of the distinct lack of underwear. Then, having watched him for a minute, I thought that he looked familiar. I then realised that he was the captain of HMS Norfolk.

In disbelief, I asked one of the navy lads that were with us, and he said, "Yeah, it's the skipper's party trick."

I struggled to reconcile the vision before me—the man that had bollocked me for not saluting was now gallivanting around the nightclub with his arse on fire! I then tried to imagine the RAF Coltishall's group captain station commander doing this, and I concluded reluctantly that it would never happen, as smashing up pianos in the officer's mess was more their bag. Maybe the navy guys were winding me up and it was another crew member that resembled their captain, because by then, I was very drunk. But my gut told me that it was the captain. Somehow, we got back to the ship in the early hours, and the next day, as we travelled by train the long way back to Norwich, I was nursing the hangover of the century. We were expected to write a report about the exchange experience and obviously none of the social aspects were included!

Just a few years ago, I had another social encounter with the lads from the Royal Navy in the UAE. Jules and I were on a long weekend break to celebrate our anniversary and we had just checked in to one of the hotels on the coast near Fujairah. The room was not ready and so we headed to the pool bar. There, we soon got chatting to about ten lads in their twenties and found out that they were submariners and that their 'boat' was in a dock, getting repairs, and so the crew were taking turns to have a few days at the hotel 'Rest and Recuperation' (or 'R&R', as it is known). The afternoon passed with lots of beers and cocktails and laughter, with them calling me a crab and me calling them fish heads. As the evening drew near, they invited us to dine with them at a local restaurant and said it would be on their shore allowances; so, basically, the UK taxpayer was going to pay for our dinner, which felt like old times to me. The meal was very nice, and later we ended up in a local nightclub.

Once again, just like the lads in Paisley, it was clear that just having a great time between themselves was the priority, despite some young ladies looking very interested. It was a bit weird being out with lads younger than my own sons, but it was a laugh as they taught me the latest dance moves and I got them doing some obviously great 'dad dancing' to a Madonna track, despite my aging hips creaking and complaining! But the undoubted highlight of the evening for us was when one of their lads fell asleep at a table near the dancefloor. The response to this naval misdemeanour was for one of the lads to fetch half lemons from the bar, which were then unceremoniously rammed into the poor sleeping lads' eyes, who naturally yelped in pain. The other lads just stood by, looking unimpressed with their errant colleague.

One of them said, "He knows the rules – no sleeping in public onshore."

To which, the victim who now had stinging red eyes and tears rolling down his cheeks, said, "Yeah, I knew the rules, I knew the rules, this is my own fault," which made Jules and I howl with laughter and that became our most memorable moment of the encounter.

One of the guys even stayed in touch with Jules on Facebook for a while. I didn't really have any view of navy lads, but I must admit from my two brief encounters with them – they certainly know how to party!

Back in Norfolk, a few weeks later, one Saturday in February, we had the neighbours' twelve and eight-year-old kids, Lindsey and Alan, around at our house. They wanted me to take them to Eccles Beach, which was about four miles away. I was down there most weekends, and sometimes during the week if I was on nightshift, as it was a great place to walk our dog, Benson. The part we visited also had a small duck pond, so we could feed the ducks before going through the massive concrete sea defence wall, with its huge metal gates, and out on to the beach.

Charts for the tide times were available, but I never looked at them as it was an exciting reveal going through the gates and up onto the beach to find out the state of the tide. Sometimes, the sea would be a hundred yards out and the beach, a flat sandy surface that was ideal for throwing the tennis ball for Benson to retrieve or playing beach cricket. The other extreme would be the waves up, almost as far as the concrete seawall defences, and you had to walk along the path instead of the beach. The weather on the day would also determine the sea's colour. The wind created waves that would crash in and send spray and mist up into the air so that you could taste the salt and smell the seaweed as you walked along. Usually, there would only be one or two other people to be seen for miles—it was a magical place and I loved it.

On this cold, breezy February winter's day, Sam had wisely stayed at home with Chris the baby, and I had Stephen in his little romper suit, booties, bobble hat and coat on my shoulders. The tide was halfway out, so we had some beach to play on and I threw Benson's tennis ball, and he chased after it as usual. Alan had brought along a light plastic football, which was about eight inches in diameter, and he was kicking that along when a gust of wind blew it into the sea. Benson's natural gun dog retrieval instincts kicked in and he dropped the tennis ball and went into the waves to get the football, but it was too big and slippery for him to hold in his mouth and so he was pushing the ball further out to sea. I was shouting at him to go to the other side of the ball to push it back to shore, but unsurprisingly, he didn't understand this and just kept trying to bite the ball and going further out from the shore.

At this point, I knew there was a problem and with hindsight, I should have just waded into the sea and got the ball back, but that window of opportunity was closed in seconds and in no time, he was already about thirty feet offshore. Now, I was shouting for him to leave the ball and come back, but he just kept trying to bite the ball, and he was probably thinking to himself, *remember the sugar beet field?* All too soon, he was over a hundred yards out to sea and going further out every second as the football was now getting even more offshore breeze on it, blowing it out to sea.

I felt helpless and was getting hoarse from shouting. We were starting to lose sight of Benson between the waves and Alan started crying, which set Stephen off. I looked at Lindsey, who was also trying not to cry.

I was in turmoil. I had two young children and a toddler in my care, miles from anywhere, and obviously there were no mobile phones then. As I looked back out to sea, I saw Benson was turning around about 150 yards out and trying to swim to the shore, he had the now punctured football in his mouth, but even from that distance, I could see that he was exhausted, and I knew that I had to do something. You read newspaper stories about people trying to rescue their dogs that have fallen through the ice on a lake and think to yourself, *if it didn't hold the weight of a dog, why did you think it would it hold the much greater weight of a person?* But now, I was being drawn by the same urge to try and save an animal.

I stripped off down to my boxers and held Lindsey's shoulders and said, "If this goes wrong, then take Alan and Stephen to the car and then try and get help from anyone you can find in those houses."

She nodded and I saw a determined look in her face as she stiffened her resolve, and then I waded out, gasping into the freezing cold North Sea. I don't think I have ever experienced anything quite as cold as that water, but I tried to put that out of my mind and dived in and started swimming as hard as I could towards Benson.

I got about fifty yards out and the exhausted dog was still about the same distance away from me, but when I shouted for him, he saw me and started swimming towards me. And he still had the bloody collapsed plastic ball in his mouth. The waves were bigger out here, and despite being a strong swimmer, it was slowing me down and I was already getting tired. I looked up and we were only about thirty yards apart now, but his eyes were wide and white with fear, and I thought I was too late as he momentarily disappeared below the surface, but then bobbed up again and was trying to swim again. In my heart, I swam as hard as I could, but in my heart, I thought I was too late, and that the kids were going to witness Benson drowning. I was also starting to wonder if I still had the energy to make it back to the shore myself.

Fortunately, for both of us, a small boat with a couple of fishermen in it had seen that Benson was in trouble and had abandoned what they were doing to come over in their boat. They hauled him into the boat and then came and got me. I was too big to get in the boat without potentially tipping it over and so I just hung gratefully on to the side as they took us back towards the shore and to the now cheering and excited kids. When we got there, I stood shivering in shallow waves and managed to get the ball out of Bensons mouth, which his teeth had by now bitten right through, and then we lifted him onto the shore. I looked at the two guys in the boat and thanked them.

They said, "We only saw the dog when you were swimming out to it."

And I said, "It was stupid of me, I know, but I had to try and save him."

To which the older boat guy said, "Boy, you would never have forgiven yourself if you hadn't tried, and especially with them kids watching you!"

He was right, of course, but if those guys had not been out fishing or doing their lobster or crab pots, then the day could have ended very differently. The kids were fussing all over Benson, who was shaking with cold and exhaustion, but then Lindsey stood up and gave me a big hug. Anyone seeing an almost naked man being hugged by an unrelated twelve-year-old girl would probably have issues today, but I knew what she meant by it. I thanked her for being so brave and strong and for looking after the boys so well. Lindsey would go on to have a career in the RAF in the very stressful job of Air Traffic Controller and I imagine that she was very good at it.

Then, the cold wind gusted and suddenly, I was even colder and shivering uncontrollably as I tried to put my clothes back on my dripping wet body. When we got home, the kids made out to Sam that I was a hero. But I knew, in a way, I had been reckless leaving the kids on their own, but there was no way that I could have just done nothing and watched my dog drown and leave the kids with such an awful childhood memory. Benson just lay motionless by the fireplace and did a little yelp when he tried to move as he had cramped out all of his legs, but later that evening, he did manage to get up for his dinner. And, then surprisingly, when he was on the way back to his spot in front of the wood burner fire, he instead came over to the settee where I was lying, dozing off, and gave me a great big lick right across my face. Despite now having the disgusting taste of Pedigree Chum in my mouth, I like to think he was saying 'thanks'. I took Benson on my own to the same beach a few days later and for Benson, it was like nothing had happened, as he chased the tennis ball and frolicked in the waves. I, on the other hand, felt like I had a dose of PTSD!

Later that month, as the world celebrated the release of Nelson Mandela after twenty-seven years' captivity, I was off on yet another training course. This time, for small arms maintenance at RAF Cosford. I had not been sent on this when I was working on small arms but now, apparently, I needed it, and so I assumed it was related to being posted back to Germany. I was not thrilled at the thought of being locked up with guns all day for a three-year tour in Germany.

This time, my experience at RAF Cosford was very different as there was no marching about, and I had my own room with ensuite in the sergeant's mess, where the food was also delicious. The course subject matter was boring to me, but fine. I decided not to drive back to Norfolk at the weekend but stay at the camp and visit relatives that I had in the nearby village of Albrighton on the Friday night.

On the Saturday afternoon, I planned to settle down to listen to the football commentaries, but after lunch, I had the sudden urge to drive to Machynlleth in Wales to visit my grandma and aunt Pam, who I had not seen since I was eight years old when I spent nearly two years living with them as a child, which was about twenty-three years earlier. Two hours later, having driven eighty miles, I remembered the five mile drive out of town to the tiny hamlet of Melinbyrhedyn.

As I crossed over the footbridge, I saw two women working in the garden before they stood up and looked at me. The old woman started walking towards me and I knew it was a much older version of my grandma.

She hugged me and said, "Now I can die a happy woman! I have prayed for this day for years."

Well, that brought a tear to my eye as my aunt joined in the three-way hug. I was given a tour of the cottage that had been a ruin when Pam moved into it twenty-five years earlier. As I had lived there for a while, I could see the changes and extensions that had been made. It all seemed much smaller than I remembered it as a child, but it was hugely impressive what they had achieved on their own—with a very low budget, some support from the kind farmer neighbours, and their own bare hands. There was now running water, so no more bathing in the adjacent stream or tin bath in front of the fire. But there was still no electricity or heating, so it was like stepping back a century. Pam was nearly forty years old and still single, but was as excited as a little girl as she showed me the articles that she had managed to get published in two women's magazines that told of how she lived in this remote place with just animals for company. She also generated a small income by writing poems and short stories and that, with grandma's pension, was what they lived on. As it got dark, the kerosene lamps and candles were lit and they told me about their life, and I explained about what had happened to me. I said that it was a shame that we had not stayed in contact and my grandma went to a cupboard and pulled out a pile of about thirty letters—birthday and Christmas cards, all addressed to me at my home in Codsall, Wolverhampton, but all marked in big letters 'Return to Sender'. She told me that she had never missed my birthday each May, or Christmas, but they had all come back to her, and it had broken her heart.

I do remember one birthday when I was about ten and I got the post as soon as it came through the letterbox in the front door, and some were birthday cards addressed to me. One was from my grandma in Wales, and it also had a ten-shilling note in it as a present. I can remember being as excited that I had a card from my grandma as much as getting the money. Later that day, I was called to the living room and my dad was there with my grandma's card and he asked to see the ten-shilling note. I went and got it. He told me that both my grandma and aunt were mad and living in a lunatic asylum, which is why I never saw them. Also, he would give me a five-pound note in exchange for the ten-shilling one, but on the condition that this was a present from him and not my grandma—*did*

I understand? I knew instinctively that this was not right, but I accepted the money as it was worth an increase of ten times. But the fact that I can remember this so clearly suggests it had an impact and that I didn't believe him, which is probably why I tried to run away to Wales twice in the following few years.

It was quite late that evening when I drove the eighty miles back to Cosford, and I was both upset and angry about what had happened. According to my grandma, there had been a big split with my father after she told him that I needed to go live with him and his new wife as I needed a father figure in my life.

Obviously reflecting on that, I don't think she was right. And apparently, he was quite happy in his new relationship without having me involved. I have no way of knowing if that was true, but his subsequent actions would suggest that there was some merit in the suggestion. I spent the following Sunday writing a long letter to my father, stating what had happened and how angry I was, and that I did not want my boys contaminated by his evilness, so he would not be seeing them again. This approach obviously meant that I was now denying him contact with his grandchildren, which was the very thing I was complaining about happening to me, but it didn't feel that way. I felt I was protecting my kids from a source of evil. I would only have contact with him twice more before a thirty-year break.

The first was a few months later before we went back to Germany and we visited family in Devon with no intention to see my parents, but my brother Nick said that they wanted to meet us and sort things out. So, reluctantly, we went to their tourist goods shop on Paignton harbour, where I found out that Nick was winging it like some sort of peace emissary when, in fact, my father was still furious over the letter. So...that visit did not go well. The other time was after the first gulf war and I returned to Germany. My father and stepmother Sylvia turned up on our doorstep. Sam did not often get really angry, but this unsolicited approach had her threatening me with divorce if I had anything to do with them. She was right, of course, and over dinner in their hotel that night, I told my parents that there would be no reconciliation by me, and that was that. I didn't have much to do with any of my family in Torquay after that, but years later, I did reconcile with my two half-brothers, despite Nick in 2022 still 'trying to do the right thing' and get a reconciliation with my father going. I think that now, he finally accepts that it will not happen. As I mentioned earlier, I don't think that my father is in any way like the awful person that he was then and I have forgiven him for what he did, especially as now he probably can't remember much of it. But there is a part of me that thinks if I did reconcile with him, he would smile and say something like, *I knew I would get you to come back to me in the end*, which would just make me furious all over again! But at least now I do have a relationship with my brothers and their families, which is nice.

In May 1990, I completed the final training course related to my EOD post by going to RAF Wittering for the Airfield Explosive Ordinance Disposal course, which basically trained me to be a light tank commander. It trained me on how to use its gun to shoot and explode enemy's unexploded bombs from a distance so that the subsequent damage craters could be filled in and the airfield made operational again. It was a lot of fun for a couple of weeks, learning skills that, even at the time, I doubted I would ever need to use.

Writing this all down has made me realise just how much I was away from home during the first two years of Stephen's life and the first year of Christopher's; and, of course, whilst I was away and often having fun, Sam was at home struggling to bring up a baby and a toddler with very little money. As I have already said, I am glad the roles were not reversed, as I would not have been able to do the great job that she did.

That June, I once again spent three days going through the Officer and Aircrew Selection Centre process of tests, problem solving and leadership exercises, and again I got through it. But, as usual, I struggled with the time vs speed, distance vs fuel use, exercise that required a good head for mental mathematic agility—something I have never had. Even when my father gave me the chance to take O-levels at South Devon Technical College when we moved to Torquay, I struggled with the math. If I am asked to remember a string of numbers, I struggle to do it and to make matters worse, I will probably transpose them in my mind after just a few seconds.

I remember my math teacher at South Devon Tech (called Professor Lamptey) had even given me individual after-hours tuition to no avail. In the end, he declared that I had the desire to do it but just not the mental ability and so he simply taught me how to pass the exam by going through the right process and showing my working out on the answer paper—as marks were awarded for the process even if the answer was wrong. The last question on the O-level paper was always about graphs and charts and apparently was considered a tough question, but I actually found that one easy as there were visual clues to work with, unlike when trying to do algebra and balancing of equations. Anyway, I passed with a 'C' grade and have been grateful for the help I got ever since.

After the problem-solving tests at OASC, they give you a debrief and the officer doing mine said, "You are not great with mental maths, are you?" I agreed that it was not my strong point, but then he said, "However, you do have great problem-solving skills."

Apparently, in the test, I had got the 'correct process' answer wrong. But on the answer paper, I had written an alternative plan as my preferred option, but I hadn't worked out the maths to see if it would work. This guy did and apparently it would have succeeded and been around the same time as the standard answer. He thought that nobody else had ever come up with that solution before. So, I thought that would give me smarty points despite getting the answer wrong. I had prepared well for the selection process this time.

I now had a genuine interest in politics, economics and current affairs and I had even sought out the supply officers at RAF Coltishall to find out everything I could about that role. That may have not been a good idea as I realised that being a supply officer would be deadly dull and I would soon be bored shitless. But it would mean a substantial pay rise and I could leave after a few years, and as an ex-officer, have better job prospects that I had at that time.

Back at Coltishall a week later, I was summoned to go to see the OC Engineering. He told me that I had failed the selection. I was disappointed as I thought I had done enough this time. He read out part of the report which determined that I had a high level of competence, political awareness but that I needed to work on my math—no shocks there. But the damning part was that through the interview process, they felt that although I was keen to be an officer, I was not genuinely interested in being a supply trade officer, despite having natural problem solving and logistical abilities, and so they failed me.

I explained that after meeting the supply officers at Coltishall, I did think I would only last a few years and that I needed excitement in my life. I told him that I was now dreading going to Germany as I was going to be working in an armoury servicing small arms for three years and not on a front-line squadron. He suggested that I wait a few years until I was eligible to apply to be a branch officer, which was designed for ground crew SNCO airmen to become officers in the trade that they had the most relevance to. For me, it would mean engineering officer, and so the chance to remain in the armament world. That was the role that I really wanted.

In July 1990, we were packing boxes, ready to move back to Germany and, for me, back to RAF Gutersloh. Also, I think OC Engineering had done me a huge favour and talked with the posting department at RAF Innsworth, as I was now not going to service small arms weapons after all, I was posted to 3 (Fighter) Squadron Harriers. I was sad after six very happy and settled years to leave the great camp that was RAF Coltishall; however, I was also excited at the prospect of being back on a front-line squadron again and having more money to enjoy all the great things that Germany had to offer.

9. The Return to Gutersloh

The first thing that struck me about working on 3 (F) Squadron was that now, unlike the Jaguar squadrons, working a nightshift starting at 4:30pm meant just that, and getting home around 3:00am was not unusual. This was because the Harrier GR5, and then GR7, were not designed with maintenance in mind – an engine change meant the entire wing had to be lifted off. Also, because they were brand-new aircraft, every new technical problem was a mystery that had to be solved and learned from, and the quick fix solutions became available for future use.

Fig 20. A GR7 Harrier of 3(F) Squadron (courtesy Maurits Even)

Within a few weeks of my arrival, I found myself out in the woods at Sennelager near Paderborn on deployment, but this time, I was in tents near the aircraft, hidden in camouflaged 'hides', where, after each sortie, we would carry out an Operational Turn Round (or OTR), which gave us about forty minutes to winch on two new fuel tanks, four 1,000lb bombs and two sidewinder missiles. Thankfully, the centreline gun on the Harrier was still under development and would prove to be a nightmare when it was introduced later. The role of the Harrier was still the same—to slow down an advance of Soviet tanks until the cavalry arrived from the USA, but we would be operating from supermarkets, motorways and anywhere that we could get fuel and weapons delivered to, until the planes did not make it back.

There was a great bunch of armourers on 3 (F) Squadron, which felt a bit like when I was on 20 Squadron as there were so many characters. I can recall Nick Pattison, Pete Stanton, Huw Jones, Graham Harvey, Mick Carrington, Alex Jackson, Budgie Burgess, Mick Holmes, Paddy Atwell, Skip McCabe, Paul (Haggy) Haggard, Colin Bradbury, Tim Kay, Simon Sutton, Ian Williams, Les Duquemin, Ashley Verncombe, Wilf Pickles, Any Thomas and Mick Smith. I was in an OTR team that was led by CT John Onions, who was a little guy married to a Thai lady called Oy, who provided an essential ingredient to our OTR team initiation ceremony, which entailed eating a complete raw chilli by chewing it for thirty seconds before swallowing it.

Oy provided the small but potent red-hot chillies. John was a kind man and for his own team, he would present a medium hot chilli for consumption, but for visitors to our tent, and especially ones that bragged about being able to eat Vindaloo curries, he produced these comparatively small one-inch-long yellow and orange chillies for them to eat whole.

I have seen several cocky guys just pop this into their mouths, laughing, and then their faces turning puce as they would shudder in agony as they chilli spread its heat around their mouth. We were all looking at our watches and one of us was doing the countdown, in which, naturally, we made thirty seconds into forty seconds, which for the guy chewing the chilli must have seemed like an hour. Eventually, we would shout 'time' and the victim would gratefully swallow the inferno inducing fruit where it would later play havoc with their stomach and insides until it was eventually dispatched in the Portaloo. The victim would always then reach for their beer in a vain attempt to douse the burning sensation in their mouth, but that just made it worse. Having a bum the next day that looked like a Japanese flag became a common reference to our OTR teams' initiation ceremony!

Of course, in the forces there is always pay-back, and John Onions got his whilst on detachment at Decimomannu in Sardinia, when the lads got him so drunk that he passed out. They took him back to the SNCO's block, next to which a new building was being constructed. Despite it being past midnight, breeze blocks were taken from the building site and used to lift John's bed a foot higher from ground level. They also blocked out the windows with thick black polythene bags and removed the lightbulbs from the ceiling and his bedside light. When John awoke in the morning with a belting hangover, he told us that he stepped out of his bed and dropped down a foot that he was obviously not expecting. He tried the lights, but they were not working, and so stumbling around his room in the dark, he eventually found the door. But when he opened it, he discovered that like a scene in a macabre Edgar Allan Poe novel, he had been bricked into his room with breeze blocks. He eventually managed to climb out of the window to escape and took it all in good humour, which is 'if you hand pranks out, then you must accept that they will happen to you.'

The pilots were not immune to having jokes played on them either. If the groundcrew were responsible for loading the aircrew baggage onto the transport plane then quite often any pornographic magazines that had been accumulated by the groundcrew would be inserted into the aircrew luggage so that they may have an embarrassing moment going through customs, or if their wives unpacked for them when they got home! Once when unusually, both aircrew and groundcrew were all staying in the same hotel in Greece, the Wing Commander senior pilot returned to his room late in the evening to find that all of his luggage had been carefully repacked into his suitcase—which must have made him wonder if he was losing the plot for a short while.

Back at Gutersloh, I was allocated a modern married quarter out at Avenwedde about ten miles from camp, which was about a twenty-five-minute drive. But by going on my mountain bike across country, I could ride it in about forty minutes. So, in the summer months, that is what I did when on day shift. Sam and the boys came over and we set up home. I got a huge Volvo 240 Estate and set up the car seats in the back cargo area so that they were facing backwards, and they both had toy steering wheels and dashboards. Benson would go in there too and so anyone following the car would see two little boys driving and a black Labrador's head in between them!

The Berlin wall had fallen by then and on 3rd October, German reunification was official, and all the internal east-west German borders were opened. Three days later, we drove through what had been the border at Helmstedt and on to Magdeburg. It was a surreal experience because, as soon as we left West Germany, the landscape changed. The roads were in a much poorer condition, horses towed ploughs across the fields, and, in the centre of Magdeburg itself, there were cobbled streets, buildings with gunfire holes, and lots of people trying to sell Soviet and East German military clothing, which, sadly, at that time I did not think was worth buying – silly me!

What was clear to us, was that reunification was going to take time and cost an awful lot of money. It is testament to the Germans that when Jules and I drove down that way from Potsdam in our motorhome a couple of years ago, there was no way of knowing that you were in what was East Germany – the roads were excellent, and the fields were plied by farmers with ultra-modern equipment. My one regret was that I never went to Berlin when the iron curtain was in place. They ran trips from Gutersloh when I was there in 1979 and I should have taken advantage back then, but I wrongly assumed it would be the way it was forever. Later, we would attach a huge five-berth caravan to the Volvo estate and drive to the Harz mountains on a regular basis, with drivers trying to overtake us, only to realise just how long the Volvo estate car and five-berth caravan was!

On 28th November 1990, John Major became Prime Minister of the UK, just in time to have to deal with Saddam Hussein's invasion of Kuwait. As the crisis deepened, it became clear that a military intervention was going to happen. During the Falkland's war, I could not go as I didn't have Harrier experience. Now, I was on a Harrier squadron, but it would be Tornadoes that would be going. But, with Sam's reluctant consent, that did not stop me from applying to go with a general application form to my SENGO. A notice to the rest of the station was put out and that resulted in about twenty volunteers, of which, three other armourers applied: Tim Kay (also from 3 (F) Squadron); and, in our OTR team, we had Pete Hogg and Darren O'Brien from the armoury. A couple of weeks later, the four of us were being driven down to RAF Bruggen for two days' intensive Tornado OTR training, which was an aircraft that none of us had ever worked on. The training was hard and after two exhausting days, we could do an OTR as a team, but we were very slow. The instructor said that we basically knew what to do and that we would get better when we did it every day out in theatre. Back at Gutersloh, we were in the medical centre as they injected us all with Anthrax, Polio and Plague, all at the same time!

A few days before Christmas, we were on a bus, and I said goodbye to Sam and my boys outside the new armoury at RAF Gutersloh. I had brand-new NBC equipment, NAPS tablets and atropine devices complete with a morphine tablet in the cap. I also had a 9mm sub-machine gun on my shoulder and two full magazines of ammunition. As I said goodbye to my family that night, I was not sure that I would ever see them again and the moment hit me – this was the real deal; I was off to war for my country. We flew through the night and arrived at Tabuk in the north-east part of the Kingdom of Saudi Arabia as the latest attachments to the 15/20 Squadron Detachment.

We were taken to our accommodation, which was in a compound at the edge of the base. But, getting to it involved about a five-mile journey through Tabuk, to the outside of the Saudi Air Force base. We all laughed when, along the way, in the middle of the desert, some wag had put up a wooden painted sign stating 'pick your own strawberries', which were up a track leading from the main road. Obviously, other than when we were camping with the Harriers, being RAF, we were used to staying in hotels; but here, I was sharing a twelve-man room with my guys in cramped accommodation in a bomb dump, and I had volunteered for this! But whilst the accommodation was basic, the sports facilities on the base were amazing and of an Olympic standard, with several swimming and separate diving pools as well as state-of-the-art gymnasiums, football, volleyball and tennis courts. But we would only be able to enjoy these facilities for three weeks before the war for us started in earnest.

We often had false air raid alerts, as when the Iraqis launched their Scud missiles towards Israel, they would initially be heading towards Tabuk until their navigation systems made them change course towards Tel Aviv. This happened one time when we were in the mess for dinner sticks in my mind. The air raid siren went off and the hundred or so guys that were there immediately dropped their knives and forks and donned their NBC facemask, shouted 'gas, gas, gas', pulled up the hood and put on the cotton gloves followed by the black rubber NBC gloves, then put their helmets on. We had no bomb shelters to go to and so everyone just sat down and looked at the meal that they now couldn't eat. There were four Philippine staff behind the food serving counter and they were now staring in horror at all of us, fully protected from a chemical attack as they just stood there, wearing their chef whites. Someone realised this and kindly handed them an NBC Facelet each—these are designed for if you are driving a vehicle as it covers your nose and mouth and gives you a few seconds' protection whilst you pull your vehicle over and stop it to get your full NBC gear on. The kitchen staff were pathetically grateful that they now had these and wore them permanently around their necks, in case the air raid siren went off—it was the ultimate placebo effect.

We were working in the bomb dump, preparing weapons and building the 1,000lb pave way bombs that we were arming with the latest British 960 Multi-function bomb fuse (MFBF). This would give our pilots to attack targets with pin point accuracy, as the bombs, when dropped and armed, would steer themselves using moveable canards on the nose of the bomb to follow a laser cone until it hit the target. These bombs could be put through the top window of a house, if that was where the target laser was directed, either from a special forces operative on the ground or from an accompanying Tornado or Buccaneer aircraft doing the spiking targeting role.

Fig 21. Myself and Tim Kay 'prepping' 1,000lb Paveway bombs (Courtesy Tim Kay)

But the focus was on doing OTRs with another weapon system called JP233, which was a British submunition delivery system that consisted of large dispenser pods fitted to the underside belly of a Tornado and used as part of a low altitude airfield attack mission to take out and then deny use of enemy runways. Inside each dispenser pod were thirty SG-357 cratering bomblets designed to blow holes in concrete runways, with a shaped charge and two hundred and fifteen HB-876 anti-personnel mines. These mines would deploy after landing to stop repair crews fixing the craters by exploding, like hand grenades, if they were approached or pushed by a bulldozer blade. To keep the enemy guessing, they would also detonate at pre-set intervals.

As part of the OTR, the massive JP233 delivery pods were often tricky to load; and, as a team, we struggled to get them on. We were doing our best, but being new to Tornadoes, we were inexperienced and slow, and the squadron guys were reluctant to spilt us up between their teams as they thought we would then have four slow teams rather than one. Armourers have a great reputation for sticking together and supporting each other, but I have to say we didn't feel the love from the Tornado squadron armourers, who, I suspect, resented us even being there. It all came to a head during a training exercise wearing our desert DP's, full NBC gear, during a sunny day of 25 degrees Celsius. There were four Tornado aircrafts being loaded and the other four teams finished a good twenty minutes before us, as we again had problems loading the JP233 pods. The Detachment Commanding Officer had watched the OTR exercise and was not impressed with our performance and he called me over to see him.

When I got there, he took me to one side, away from everyone, and said, "Look, this is not your fault. You are not Tornado guys, but you are just not good enough to be in a live war environment. I knew it was a bad idea bringing you out here and now I'm going to send you all back to Germany."

I was feeling a mixture of disappointed and angry, and I felt that we had let the entire Harrier force down, but I was not giving up easily and said, "Look, sir. It's true we are not Tornado guys, we only had two days of training at Bruggen, and, to be honest, since we have been here, we have had very little support from the squadron guys, who seem quite happy for us to struggle. So, give us a chance to prove ourselves by giving us proper training and then, if we still can't do it, then fine, send us home. But don't send us now without a fair crack of the whip."

He went silent for a moment as he considered things and said, "Okay, Sarge. You have three days before we do this exercise again and that is it. Fair enough?"

I agreed it was and he instructed one of the Weapon Training Cell SNCO's, I think it was Tony Dunk, to personally train us. Tony turned out to be a great trainer as rather than just do the entire OTR, we practiced each component individually until we got it dead right. He gave us tips to save time and showed us little tricks of the trade regarding loading the tricky guns and the dreaded JP233 dispensers.

On the second day of the training, we just did back-to-back OTRs all day and into the early evening; by the last load, despite all of us being knackered and wearing our full NBC gear, we had the OTR load time down an acceptable range. We were now exhausted, but we were ready.

The following day, we were back on the squadron dispersal area with the other four aircraft load teams, ready to go again. The Tornado guys were no doubt really pissed off that they were having to do the OTR exercise again. Ignoring them, I fired up my guys by telling them that we needed to not only do a good OTR time, but also needed to beat one of these Tornado motherf*cker teams to really make a point. But I also stressed not to rush, just move quickly so that we would be fine. I saw them all stiffen up, ready for the challenge and off we went! We did the perfect OTR load, using the tricks we had been taught, and even the tricky JP233 dispenser pods went straight on for once. I just focussed on what we were doing, and as I finished setting the weapon load computer and stepped back, the guys shouted out "tools all accounted for," and we took off our gas masks and tried to cool our sweaty faces. Only then did I look at the other OTR teams. One other squadron team had just finished too, but I was amazed to see that the other two teams were still going at it. We had beaten two of the Tornado squadron OTR teams!

I looked across and saw Tony Dunk, who was one of the official observers. He was smiling and gave me a discreet 'thumbs-up' sign down by the side of his leg and I was delighted. The other Harrier team guys, now realising what we had achieved, were beaming from ear to ear; the Tornado team guys that had finished, not so much. In fact, they looked quite embarrassed. The exercise finished and all the teams had posted acceptable OTR times, but there was no doubting who the real winners were. To be fair to the Tonka boys, some of them did come over to congratulate us, but the real reward was the Detachment Commanding Officer calling me over to him again.

He said, "Sarge, did you know that only the top two percent of RAF pilots are selected to fly Harriers?"

I replied that I didn't know that, but that I wasn't surprised as they had no navigator and had to fly and do everything themselves.

To which, he said, "Exactly. Well, from what you guys have achieved here today, it would appear that the Harrier force have the best armourers too. Well done, Sarge. I didn't think you would pull it off, but you have all shown everyone that you chaps deserve to be here."

I thanked him and walked away feeling like Tom Cruise did at the end of the film 'Top Gun' on the aircraft carrier, when he had saved the day.

Fig 22. JP233 On their loading trolleys (Courtesy Tim Kay)

Fig 23. JP233 pods loaded onto the Tornado (Courtesy Tim Kay)

Christmas day was a strange affair. There was a Christmas dinner in the mess and the RAF had smuggled in some beer. We were each given two small tins of Amstel that we had to drink and then return the empty tin so that it could be crushed and flown back out of Saudi on the next Hercules. I didn't fancy just having two small cans of beer and so I gave mine away. We were also allowed one five-minute telephone call home, which was tough – talking to the boys and not being with them on such a special day.

I had wanted Sam to stay at Gutersloh so that she had a support network and would be kept informed, but the only other guys deployed from Gutersloh were the 18 Squadron Chinook helicopter boys, and it seems that their wives were not very interested in anyone else; so, she had gone back to the UK and stayed with her parents.

The first gulf war started properly for us on Thursday, 17 January 1991. That night, we launched most of our Tornadoes armed with JP233 so they could do the very risky low-level attack on various Iraqi military airfields and civil airports. I am not a religious guy, but as I watched the jets climb and switch off their anti-collision lights once the engine's afterburners stopped and fly away, I did say a little prayer for them to all return safely.

That night, they did all return and after landing, the pilots were all so excited about their mission, they even wanted the brass connecting devices that had held the JP233 dispensers to the airframe before they were jettisoned. One came back with an empty dispenser that had failed to jettison and it looked strange with all the ports exposed, with the metal peeled back after the bomblets had burst through the skin of the container.

We heard later that one of the pilots attacking Bagdad Airport had lined up for his runway attack. And, only at the last minute, noticed that there was a Russian Mig fighter about to take off. It never made it as the JP233 bomblets just destroyed it in-situ and that Tornado became known as the 'Mig Killer'. As I recall, we only used JP233 for a few more sorties and then the remnants of the Iraqi Air Force flew in a mad dash to seek haven in Iran, who previously had been their deadly enemy. There was then a switch to dumb 1,000lb bombing of airfield Hardened Aircraft Shelters (HAS) and other strategic military targets through bombing missions. But when a Buccaneer aircraft arrived to provide Paveway spiking, with its Thermal Imaging Airborne Laser Designator (TIALD_ capability, we then switched to Paveway laser guided bombs. As an armourer, it was an unusual experience to not only build the bombs but then also go and load them onto the Tornado aircraft that flew off that night and dropped them; it had a very satisfying feel about what we were there for. But a few nights later, we lost a Tornado out of Tabuk, and, at the time, they were not sure if it had been shot down or blown up by its own bomb. It was thought to be the latter, in that, when the bombs were released from the aircraft, the laser detector of the second bomb locked on to the rotating arming vane on the tail unit of the first bomb; and, due to the strobe effect, it appeared stationary, so the bomb thought it was the ground and promptly detonated. As the weapons were still relatively close to the aircraft, the explosion brought it down, killing the aircrew at the same time.

The next day, the boffins in the UK had decided that the fuse set to 'airburst' may have been the problem, and so a different arming code for the 960 fuse was needed and the instruction came out that night.

The morning after, I was called in to see the Squadron Leader OC Engineering and I was instructed to go alone around all the squadron dispersals and personally change the 960 code on each bomb. So, armed with a screwdriver and the code setting box strapped around my neck, I spent about six hours walking around the various dispersals, changing the fuse code, and recording it on the weapon log sheet.

We had happened to be on nightshift when the war started and the powers that be decided that they would keep it that way for several weeks. As a result, one shift, we were night owls and seeing all of the action with the sorties taking place; the day shift was just repairing the aircraft and loading them ready for the next night. As they were often not busy, they could sunbathe and relax whilst we worked solidly each night for twelve hours. After five weeks, one shift looked like the cast from the 'Walking Dead' and the other, all brown and tanned as if they had been on 'Love Island', not that either show existed back then.

Finally, they changed the shifts back to one week on nights and then swapped to a normal week of days, then swapped to nights, which was much better. We settled into a rhythm of bombing sorties each night and preparing weapons by day. It soon became apparent that the Iraqis were no real threat other than firing off their unreliable SCUD rockets, and even that threat diminished as their launch vehicles were sought out and destroyed. The routine of working nights and its daily bombing sorties carried on for a month and it soon became clear that the Iraqis were not going to be able to attack us, other than with random SCUD missile attacks.

The one time we did briefly think the Iraqis were attacking us was on the night of 7th February. We were on the flight line, when the air raid sirens went off and everyone scrambled off to the air raid bunkers. I never did this. I just walked off the pan and found a comfortable spot to lie down and watch the action that never happened. But this time, there was something happening, as streaks of light streamed over the sky towards us. Some guys were panicking and donning their gas masks, shouting 'gas, gas, gas', but I could see that the streaks were way too high in the sky to land on us; they soon passed by and disappeared over the horizon. It turned out that they were indeed Soviet manufactured, but in fact, it was the Salyut 7 space station breaking up as it re-entered the earth's atmosphere.

We prepared and dropped that many bombs that we actually ran out of the new 960 fuses. When the next batch arrived, the manufacture date was only two weeks before and the explosive filling date was three days earlier! To supplement the modern fuses, we were also sent out a supply of old mechanical pistol fuses and fulminate mercury detonators along with manuals that had 'Obsolete' stamped across every page in large capital letters. Luckily, I had seen these things before during my weapon mechanics training, when I first joined up, but it was surreal to be jumping between state-of-the-art technology fuses and then World War II mechanical ones!

What I did like about how we were operating in the war was the complete absence of bullshit and the flattened chain of command. Our squadron leader OC Armament boss, Graham Cooke, would attend daily command centre briefings. In the evenings, he would visit the bomb dump where we were operating from, and at shift change, give us all an intelligence briefing and update on what was being planned. This was great as we were getting the information fresh and first hand as opposed to normal times, when, by the time the message got down to the guys on the ground as a written directive, each level of command would have made the message their own to make them look important.

This would inevitably lead to nuances and different interpretations being made so that often the message that we finally got, was far from the original, sometimes unworkable, and made the senior policy officers look incompetent and cause a backlash until things were resolved. I remember the boss bringing a video to the bomb dump to show us after his daily briefing and it started off with the same music as at the start of the film 'Top Gun' with the 'dong' noises and electric guitars of Kenny Loggins 'Danger Zone' playing as the aircraft prepared to launch. But then, the video cut to an ariel view of a HAS being targeted by the TIALD laser with the crosshairs on the middle of the roof of a HAS, and then the aircrew commentary saying, 'Bomb gone'; then, about twenty seconds later, there was a silent explosion. As the blast wave and dust cleared, you could only just see what was left of the HAS and some background music by Berlin from the Top Gun film 'Take My Breath Away' played, which obviously someone thought was amusing. The crosshairs then moved across the ground to the next HAS and the same thing happened. There was another ten minutes footage of this, but I had left the room by then.

Hardened aircraft shelters had been built in the 1980s as the best way to protect aircraft and the personnel that operated and maintained them. I had spent many exercises sat inside one, playing war games, but here was the hard evidence that they were an utter waste of time and just a target if the enemy had accurate bombing systems and concrete penetrating bombs. As each HAS was destroyed, I could not shake the thought that inside each one were possibly some Iraqi guys just like me, just doing their duty and that there was probably no enemy aircraft in there anyway as they had all flown off to Iran!

I kept my own council, but I was very uncomfortable with this unnecessary glorification of war, when really, it was just like shooting fish in a barrel. By the end of the war, somehow Graham Cooke had found time to paint a scene of some Tornadoes sat on the pan outside a HAS in Tabuk, ready for its next sortie. He called the painting 'Minutes to go' and when he had finished it, he had a hundred or so prints made, he then personally signed each one, and gave us one each as a gift and a 'thank you' for our efforts that he said had resulted in him being awarded an MBE. We all doubted that we were responsible for him getting the honour, but the personalised painting print gift was a really nice touch and mine proudly hangs framed in our kitchen today. He also kindly agreed to me using the image on the front cover of this book.

Fig 24: ‘A few minutes to go’ painting by Sqn Ldr Graham Cook, OBE

I had a dual role on the detachment—I was a weapons preparation SNCO and the RAF’s EOD representative. There was a suspect package sent to one of the squadron’s mail rooms that I had to deal with. With my No. 2, we prepared the Pigstick Disruptor, which was basically a piece of tubular equipment that injected 100ml of water at high velocity into an IED in order to dynamically separate the various components of an Improvised Explosive Device (IED) before it had chance to detonate, and, so, render it inoperable. As in all the EOD films, I walked in alone past the cordon tape and entered the building carrying the pigstick and trailing the firing cable as I went. I entered the mail room, saw the package on the floor and was focussed entirely on setting up the pigstick to produce maximum disruption to the package after I had left. As you can imagine, I was slightly tense and very focussed as I had no idea if this was a real device that could go off at any moment.

So, the last thing I needed was to suddenly find an administration officer next to me saying, “So, Sarge, what do you think it is?”

I nearly jumped out of my skin and shouted at him to, “F*ck off and get out of here.”

He looked shocked but got the message and then left immediately. I recomposed myself and finished the job before leaving the building. I saw the admin officer and shook my head, walked over to him and told him that he was lucky I hadn't punched him in the face for startling me like that. I asked him what the hell was he doing going in past a safety area cordon. To be fair, he looked sheepish and apologised, so I said I was sorry for telling him to f*ck off and that was that. My EOD number two pressed the firing button on the shrike and there was a muffled sound from inside the building. I went back in to see what was left of the suspect package and found no devious terrorist explosive bomb components, but instead, a load of now shredded children's drawings that a school in the UK had sent out to cheer us up. And that was my one and only 'long lonely walk' to a suspect IED device of my entire RAF EOD career!

Whilst my EOD exploits are nothing special, I suspect that our little EOD team of Harrier boys still hold the unofficial world record for creating the largest deliberate man-made explosion with munitions in the world! We were tasked with getting rid of a load of ordinance that was either defective or past its life-expired date. The brief to me was very simple—I was to take a lorry of high explosive ordinance, drive out in a southern direction to anywhere I liked in the Saudi desert that was miles from anywhere and then blow it all up and get back to base before nightfall.

So, that afternoon, we loaded the lorry. I can't remember the exact numbers, but as you can see from the photographs in this book, there were at least two US-made 2,000lb bombs, two UK-made 1,000lbs, about 500 × rounds of 20mm High Explosive ammunition and a bunch of other ancillary explosive items. The back of the four-tonner looked like a terrorist's wet dream.

Fig 25 & 26. Sgt Paul Hilton carefully unloading the High Explosive bombs from the truck (Courtesy Tim Kay)

The next morning, I collected the Saudi authorisation papers to show any Saudi police that stopped us, and we set off in three vehicles and drove for about two hours south of Tabuk, out into the desert. We were stopped at a checkpoint only once. I don't know who signed the Saudi orders, but the police were very impressed and even gave me a smart salute as we drove off. All signs of human habitation petered out apart from the odd camel farm and so we turned off onto a track heading up into some hills. I should explain here that most people associate the word 'desert' with an image of rolling sand dunes; and, whilst areas like that do exist, in Saudi, most of the desert is a mixture of hard sand with stones and rocks in it. It's like driving over a rough road.

As we climbed one hill, we came across a bizarre sight of the remains of several, what looked like World War I, biplanes. They were just sat there, miles from anywhere, preserved in the dry desert air but we didn't have time to linger as in the gulf. The sun would go down in just a few hours and we had lots to do. We located a spot in a valley that I liked as we would be able to retire a mile or so back up the hill to observe our explosions and be able to see every approach for miles around.

We unloaded the lorry and created this huge pile of explosive ordnance, but as a trainer, I also wanted to get some training value out of the day and so we also separated two 1,000lb bombs for some training techniques.

Fig 27: Several thousand pounds of explosive ready to go bang! (Courtesy Tim Kay)

We had one of the US EOD team along with us as an observer and he was busy filming everything with his cine camera. With my number two, we prepared plenty of PE4 plastic explosives, detonators and detonator cord to make sure that the entire pile would explode simultaneously, as I feared munitions being thrown out of the explosion. We would then have to deal with it afterwards and the time was pressing. I was measuring out the safety cord that provided a burning time delay, as you see in films like 'Mission Impossible', and from the tables, I calculated the length needed to give us thirty minutes and then started measuring it out by taking it in my hand from the cable spool until it touched the tip of my nose, which gave me a yard.

The US EOD asked me what the difference between a yard and a meter was and so I showed him that if I turned my head away, so my nose was further away for the cable to reach, then told him that was a meter; he understandably looked very dubious!

The RAF EOD guys at RAF Wittering had been trying to devise ways of splitting open the thick iron cases of high explosive bombs without detonating the explosives inside and they had developed a small, shaped charge called an X1E1. It had a copper cone inside it, and when about an ounce of PE4 plastic explosive and a detonator was added, it formed a shaped charge that would fire a plasma jet at the bomb casing when detonated. They had sent some of these devices in case we needed them, and here was a chance to try one out. The trick was getting the height of the shaped charge stuck to the bomb at the right height by bending the wire legs and taping them to the bomb casing.

Fig 28. X1E1 – One ounce of explosive PE4 to safely disarm a 1,000lb bomb? (Courtesy Tim Kay)

I prepared it and then added safety fuse, this time with a twenty-five-minute time delay, as I needed the training explosions to go off before the motherlode explosive pile, in case that sympathetically detonated it. The other technique those ingenious EOD guys at Wittering had dreamt up was called 'Round Tom.' It used the detonator cord and plastic explosive to 'cut off' the back of a 1,000lb when the detonator cord detonated, and also detached the weapon's fuse in its holder pocket before it could detonate; so, the result should be a small explosion and much less damage, rather than have a crater and blast if the bomb detonated in 'high order', as designed.

So, on the other 1,000lb bomb placed about a hundred feet away, Paul Hilton set up the detonator cord and PE4 and the US EOD guy captured it all on film. This time, I calculated a twenty-minute delay on the safety fuse. One last check, then we got drivers into the two vehicles, and I made sure they had engines running, and that the other two EOD guys had moved into position before I gave the signal. All three of us lit both the safety fuse on each setup before we ran to the vehicles and got the hell out of there, just in case the safety fuse was faulty and burned too fast.

Back up on top of the hill, I looked through binoculars and was relieved to see three tiny wisps of smoke as the safety cord was burning merrily away towards detonating the three explosions. I made a pretentious act of doing a countdown on my watch, which was crazy, as measuring a safety fuse is an extremely imprecise art and it had varied burn times, though I had tested a strip. Anyway, I counted down for a laugh with low expectations and the US EOD guy still had that dubious look on his face. '5, 4, 3, 2, 1…' And to my surprise, the first device explosion happened right on cue. All we saw was a flash and a few seconds later, a bang, but no big explosion, which meant that the device had not detonated the high explosive and so the technique may have worked. I looked at the US EOD guy and he wasn't even filming it, but he did now at least look quite impressed.

Five minutes later, but sadly a few seconds late, the second bomb was hit by the exploding detonator cord and PE4 combination and again, just a flash and no high order explosion. So, another potential success. This time, the US EOD guy had filmed it as he was becoming a new convert to my safety fuse timing techniques.

Then, five minutes later, and bang on time, came the biggest explosion I have ever seen. The flash was enormous, even from a mile away, and a few seconds later, the very loud bang reached us and then the echo reverberated around the hills and valleys. When you think that in the UK, we are only allowed to set off one 1,000lb in high order fashion and that towns for miles away from the approved Essex coast EOD range must be warned about the bang. If our explosion had been set off there, then people in Leeds would have needed to have been warned!

A huge pall of black and grey smoke climbed into the afternoon sky as we drove back down to the ground zero of our efforts and what greeted us was amazing. Of the huge pile of explosives, there was absolutely nothing left except a pretty impressive smouldering crater in the rock-filled desert ground. Every piece of ordnance had been vapourised along with the teddy bear that Tim Kay had placed on the top of it.

Fig 29. The results of the main explosion in the rocky ground (Courtesy Tim Kay)

We drove to the X1E1 test site and the 1,000lb bomb casing lay there empty, like a peeled banana, and somehow the entire bomb shaped high explosive content charge had been lifted out of the casing and sat a couple of feet from the bomb—it was a perfect execution of the technique, except that ideally, the explosive contents would have been deflagrated all over the place.

Figs 30 & 31. X1E1 result – bomb casing opened successfully without a high order detonation (Courtesy Tim Kay)

By now, the US EOD observer was whooping and cooing in appreciation and filming away as we approached the other test with the detonator cord and PE4 plastic explosive mix. Unfortunately, the back of the bomb had not been detached from the bomb, it had just cracked the base plate, leaving the rest of the high explosive inside the casing. US EOD man was filming away whilst asking questions about the X1E1 as he was seriously impressed, and so was I, but I knew that it was also a fair bit of luck that both techniques had worked.

Indeed, the X1E1 was soon to be replaced with a ballistic disc shaped charge called a Baldrick, but still, it felt great that we had pulled it off. By now, it was late afternoon, and so we prepared one more demolition to get rid of the remains of the last 1,000lb test bomb and the explosive content from the other. We blew that up and drove back down for a final safety check before heading back along our own vehicle tracks. We made it to the highway just before darkness fell.

A few months later, when I was back in Germany, I was contacted by someone in the US EOD HQ after they had seen the footage. He was asking lots of questions as they were clearly impressed with what we Brits had developed. I just gave them the contact details of the EOD guys at RAF Wittering and I never heard anything about it again. I suspect that without seeing the cinefilm footage, even the Wittering EOD guys would not believe what we had done, and I was not going to tell them as I obviously had totally ignored all the standard demolition procedures by blowing up so much stuff in one go. But we were at war and now, thirty years on, I can talk about it; luckily, Tim Kay was a keen photographer and so the evidence is there for all to see.

The last two months of our war consisted of far fewer bombing sorties, and it was obvious that the war would be over soon, so we were preparing to pull out. My days were spent supervising repacking of 600lb anti-tank cluster bombs that were never needed and dismantling 1,000lb bombs once the hostilities ended. In March, it was all over, and we flew back to RAF Laarbruch to a hero's welcome for the Tonka boys, but those of us from RAF Gutersloh were just packed off quietly into an RAF bus for the journey north to no welcome of any sort at Gutersloh.

When I eventually got back to my married quarter, I found that the front door had been decorated by Sam and the boys, with a sign saying, 'Welcome Home Daddy' and balloons, which obviously brought a tear to my eye. To make things worse, 3 (F) Squadron were heading off to Las Vegas on Red Flag a few weeks later, and the management had decided that as we had been away for so long that we would not want to be gone for another five weeks. I was not too bothered, as I had been twice to Red Flag already with the Jaguars, but I did feel for the other guys not being rewarded for volunteering to go to war. Meanwhile, all the armourers that hadn't volunteered for the gulf went to Vegas and later, we found out about how they had all pretended to have been on 'Desert Storm' to get show and theme park discounts and free drinks everywhere they went! Once again, 'if you can't take a joke, then you shouldn't have joined up!'

Life soon returned to normal back on the squadron and when, on 18th May 1991, Helen Sharman was the first British astronaut into space as part of Project Juno, I was off back to the UK to do my Harrier GR5/7 training course at RAF Wittering. I found the course to be easy as, apart from my time in the gulf, I had already been learning on the job for several months. However, attending this course would have a profound impact on my life as this was where I was introduced to the Open University.

Like most people, I had stumbled across OU programmes on BBC2 at odd hours of the day. Depending on the topic, I found them either interesting or daunting and I had concluded that getting a degree was something other people did, not me. But the course instructor was an armourer and he told me that he was halfway through his six-year degree and that if he could do it, then anyone could do it. He explained that you first did a foundation course and could choose from several disciplines, and that taught you how to learn at that level. He suggested that I do one of those and that if I didn't pass, then at least I had tried; but he was very confident that I would pass. The best thing was that the RAF would pay for all of it, including the mandatory week's summer school and mileage to any tutorials put on in my area. So, what did I have to lose?

That summer, I investigated the OU, spoke with the education team at Gutersloh, and applied to the RAF to fund it, which was approved. In early December, I was enrolled onto the 'Living with Technology' foundation course, and in February, the learning started in earnest. I really enjoyed the course, and the next November, I passed the exams. I had the option to do another foundation course that would count towards my degree credits and was advised to do one on social studies, which was more about people and global issues. I hated it. The Technology Foundation course dealt in facts and figures leading to concrete answers, but the social studies one was much vaguer and often had statements such as 'there is no right or wrong answer on this topic.' As an engineering type, this was most unsatisfactory; but, in the end, I did realise that there was great value in trying to look at something from a different perspective and deal with viewpoints that I fundamentally disagreed with and that these new skills would help me greatly in business years later.

After the second foundation course, I spent the next four years doing courses on design and innovation and then focussed on systems processes and management and, eventually, six years later, I got a degree that would open job opportunities that simply would not have been available to me without a degree. Of course, it did mean that for ten months of the year, I had to find twenty or so hours a week to study as well as do my day job, raise a family and have a life. But once I adapted, I found myself looking forward to learning new materials except for the math and chemistry modules in the first foundation course, which, naturally, with my renowned mathematical disability, were a struggle for me.

Another learning experience came my way when I was selected to be an HGV driver for the squadron. Off I went to RAF Bruggen to drive a three-tonner lorry around the Monchengladbach area for a couple of weeks, which was great fun. I passed the test in the DAF lorry, which was like driving a big van and was easy to drive, but then they gave us training on the lorries I would be driving—an ancient Bedford truck that crawled along by comparison had gear changes that needed to be made by double-declutching between gear changes, which took a bit of getting used to.

Then, because of an EU law change, towing a trailer on the back of a lorry became considered as being an articulated vehicle and the driver had to have an HGV One licence. So, two weeks after passing my HGV Three, I was back at Bruggen. But this time, I was driving a gigantic articulated lorry, which was a very different proposition and I really struggled with all the gear changes that involved eighteen forward and four reverse gears, as well as the extreme length of the vehicle.

After a week, I was taking my test and this time, I was not confident at all. But once again, lady luck smiled on me. Into the cab jumped the examiner who was a RAF MT guy I had never met before, but who was convinced that he not only knew me, but that I had played football for RAF Gutersloh with him when he was there. He even reminded me about a goal I hadn't scored. Anyway, I did nothing to dissuade him that I was not his past football buddy, and he spent most of the time talking about people I didn't know and games I never played in with me just nodding and adding well-timed 'ah ha's'.

Luckily, I did know some of the Gutersloh first team players and so I sprinkled their names into the ongoing conversation. The test route included a very tight road through a long village with lots of parked cars and, as we approached it, I was filled with dread. I had not been great going through here in during the training. But, as we got closer, there was a traffic jam behind a marching band going down the road.

Seeing this, the examiner said, "Oh right, no worries. PJ, turn right up here and head back to Bruggen on the autobahn. I can see that you can drive this thing."

I couldn't believe my luck! Back at the test centre, I did manage to do the manoeuvring tests, such as reversing into a bay lined by plastic bollards and that was that. I was now officially a legally UK-licenced HGV One driver. Fortunately, for road users everywhere, I never drove an articulated lorry again and you will no doubt be relieved to know that my HGV One licence has now expired!

I ran the 3 (F) Squadron team as player/manager in the inter-section league and then I was asked to take over the RAF Gutersloh second team as manager only. I was nowhere near good enough to play at that level when we played in the local Gutersloh town league. We played against German teams who were about the same standard as us and several Turkish teams who liked to play with a lighter ball and do fancy tricks. They didn't appreciate the British style of physical contact football or having to play with a regulation weight football. Every tackle, they would be screaming out in pain and writhing around in agony, until they got a free kick and they suddenly popped back up onto their feet and carried on as if nothing had happened—just the way professional footballers do today!

My abiding memories are getting to runners-up in an international tournament played in Holland, despite only having ten fit players and constantly trying to get one of our armourers Colin Bradbury, who was a big lad and had pace to burn and finished well. But getting him to stay onside by timing his runs better was impossible. He recently told me that he fixed the problem by playing in defence!

My other main memory of Colin was on an RAF bus carrying the squadron families back from a squadron BBQ and pig roast. As we were about to set off, a very drunk Colin boarded the bus with the pig's head held on top of his own; and, as he ran down the aisle of the bus, the women and children were screaming. I can imagine that there are a number of those kids that still have PTSD and can't eat bacon!

Fig 32. The infamous Pig's head (Courtesy Colin Bradbury)

I can remember a few memorable games, but my favourite was when our best player, a Scottish midfielder called Callum McRobbie, turned up late for a match against one of the league's top teams. I had a rule that if you were late for match day, then you started on the bench. When I read out the team and Callum was on the bench, the rest of the team looked shocked. By half-time, we were three-nil down and everyone was looking at me to bring Callum on, but I didn't. The second half started, and ten minutes later, I called Callum over.

I said, "Whose fault is it that we are losing?" He said it was his and that he wanted to put that right. To which, I said, "Make me proud of you instead of pissed off with you."

He nodded and on he went! The mood in the team was transformed as his silky skills came to the fore. We scored three times and so were level and all over them, and now they were now hanging on for a draw. I really don't know who scored the winner in the last minute, but I will say here that Colin Bradbury stayed onside for once, timed his run to perfection and hit a twenty-five-yard screamer into the top-left corner. The lads were rightly ecstatic at the final whistle and Callum got my 'man of the match' award, despite only playing for thirty-five minutes. I will admit now that, at the time, when we were losing, I was not sure if my stance on discipline was the right thing to do, but I got away with it as we won, so clearly it was. Callum's exploits were the talk of the football club bar and I lost him to the first team, which was bad for us. But great for him and he deserved his chance.

Fig 33. RAF Gutersloh 2nd Team (Courtesy Colin Bradbury)

With the arrival of 1992, the Maastricht treaty created the European Union, and as one war-prevention political construct was created, another broke down with the start of the Bosnian war. For me, it was a good year for medals as I was finally given one for my exploits in the first gulf war and then I got a medal for fifteen years of narrow escapes and undiscovered crimes as the station commander pinned a Long Service & Good Conduct medal onto my chest. One charge and I would not have been awarded this, so it was quite something to get it!

I also had a job change on the squadron as I became one of the shift line manager SNCOs. The job entailed overseeing all the 'liney' FLMs and acting as the go between the aircrew operations and ground crew. Each afternoon, the next day's flying program would be issued. We would prepare the aircraft with the correct fuel loads and armament configuration, and the next morning, service the aircraft ready to go after the armourers had loaded the aircraft. The flying program changed constantly due to the weather, operational requirements, and aircraft serviceability, and so the job was managing this constantly-changing situation. It was stressful and needed good planning skills and I loved every minute of it.

Most SNCO's reluctantly did a six-month stint and could not wait to get off the job, but everyone seemed to like having me do the job and I ended up doing over two years on the desk. I did so long that on nightshift, I also had to stay later and do some armament trade work for a few hours, in case I forgot how to do my real job! Due to a low-flying ban in Germany, the pilots increasingly had to fly over to the UK, and so, there was an increase to UK RAF bases for two-week detachments of a limited support crew. Engine guy Billy Connelly and I were also the advance team for most UK detachments, and armed with a RAF credit card. We would be sent ahead to book off-base accommodation if required; arrange aircraft fuel deliveries; armament supplies; and most importantly, get in the T-bar supplies, especially beer, build a BBQ out of oil barrels and angle iron, and identify social hotspot venues to brief the lads on when they arrived. We must have set up at least ten such detachments and it was always a fun job.

On one not so much fun Harrier deployment exercise camping trip, we were deployed up at Bergen Hohne, where we were not allowed to dig gun pits as we may accidentally unearth unmarked graves. This area had been a notorious Bergen-Belsen concentration camp in World War II, where thousands of Jews of various nationalities had been processed and lost their lives. The joy of not having the arduous job of digging trenches was soon tempered by a visit to the little holocaust museum and watching the film footage of the awful history of the place. It was a major TACEVAL exercise, complete with NATO observers, and was being run to a strict timetable of events that created a war scenario.

One such exercise event was an unexploded Soviet bomb that was supposed to deny access to some Harrier hides for several hours until the EOD team arrived to deal with it. But I was EOD trained and so I went over to the bomb, realised that I could disable the fuse with bodge tape, and did just that. The lead directing staff officer was not happy, but the EOD directing staff said that the bomb was now safe. As I walked away, the officer called me over and told me that I had just ruined an entire day's exercise planning, as this scenario was supposed to have serious implications for the whole site and that it should have not been resolved until the early hours.

I just smiled and said, "Sorry about that, sir," without any real conviction and thought to myself, *Harrier force guys, you all owe PJ a beer!*

The next day of the exercise, I heard people shouting my name and soon I was back in front of the same DS officer. He handed me a card, informing me that I would be seriously injured by shrapnel from a bomb exploding during the next attack.

He said, "I could have just had you killed, of course, but then you would be out of the exercise and sleeping in your tent. This way, you can have fun with the medics for a few hours."

I had to smile as he had got me back with a good one. The attack happened and I had five hours sat in the medical tent, bored shitless.

I don't recall exactly when this happened and I will not mention the squadron or names of those involved. The most senior Sergeant on our armament shift (and so the de facto boss) was not always the most proactive; he liked to issue work instructions to the team before announcing that he would 'hold the fort' and man the telephone in the office whilst we did the aircraft work. If we were lucky, he would raise the paperwork. I was unaware that the lads were taking their revenge, as when he frequently requested them to make the tea and coffee, they were giving the rim of his mug special attention with their penis. But later, they took this to a new level by putting laxative chocolate instead of sugar into his coffee.

God knows how he did not taste this, especially as they kept increasing the dose in place of sugar. But one night, they had gone too far, and he collapsed and was whisked off to the medical centre, and possibly to the hospital after that. I thought that maybe he had suffered a heart attack as he was overweight and unfit, but then two of the lads called me outside and told me what they had been up to. Naturally, with my massive concern for my colleague sergeant's health, I immediately asked them if they had been 'rimming' my mug and giving me laxative too. They denied it, but that they had put half a laxative bar into his last coffee before because it just wasn't having the desired effect. And now, it was them that were shitting themselves, wondering if they had now killed him. I told them not to say anything to anyone until things panned out, but if something bad did happen, they would have to confess. Luckily, for the Sergeant and the guys, he made a quick and full recovery, and nothing was ever said about it again.

I had yet another paranormal experience at Gutersloh to challenge my scientific beliefs, but this time, not to me personally. One of the squadron electricians (whose name escapes me, but who was just an average guy, who was very serious and definitely not prone to making jokes) came bursting into the flight line hut one night, covered in sweat and looking very pale.

The shift Flight Sergeant said, "You look like you have just seen a ghost!"

To which, he stammered out, whilst trying to catch his breath, "I have."

When he calmed down, he explained that he had been working on one of the Harriers inside the undercarriage bay and he needed the battery power turning on in the cockpit so that he could complete his electrical tests. He said he knew someone was in the cockpit and called out for them to switch on the battery power for him but got no answer. He shouted again, but was again met with silence. Annoyed, he climbed up the cockpit ladder, ready to have a go at whoever was sat in the cockpit for ignoring him, only to see a pilot wearing a World War II leather jacket and helmet with goggles, staring straight ahead. The electrician fell back down the ladder in shock and sprinted the 500 yards back to the line hut.

Nobody would have believed the electrician's story except for two things—firstly, the guy was just not the type to make things up; and secondly, it turned out that he was unaware that at the back of the HAS he was in, there was a gravestone of a Polish aviator killed during World War II, who maybe have been a POW. Nobody else had experienced anything similar in that HAS during my time on 3 (F) Squadron, but it's hard to think he did not see something.

10. Moving to Laarbruch

It had been announced that the squadron was moving lock, stock and barrel to our new home at RAF Laarbruch and we had a manic period of packing up the squadron in Gutersloh and moving everything to Laarbruch whilst still maintaining operational effectiveness by flying sorties. It was like one massive field deployment and I ended up making several journeys with a three-tonne lorry and trailer to move equipment to our new home. As an SNCO, you were given pieces of RAF equipment that you had on your inventory and were responsible for. Bizarrely, I had some 1200 litre aircraft fuel tanks on my inventory that all had unique serial numbers, and one of my tanks was missing. I tried to get it 'written off' as lost from when one of the Harriers had crashed, but the suppliers told me that despite the aircraft only being able to carry three external fuel tanks, eight had already been assigned to the lost aircraft! Eventually, another fuel tank turned up, covered in brambles and nettles behind a HAS at Gutersloh. As it was no longer serviceable, it was assigned as mine, even though the serial numbers did not match!

RAF Laarbruch is in the Lower Rhine region of Germany on the border with the Netherlands. Originally built as an advanced landing ground by the British army in World War II, it became a proper RAF Germany base in 1954, and was home to McDonnell Douglas F-4 Phantoms, then Jaguars and Buccaneers, until replaced by four squadrons of Panavia Tornado (2, 15, 16 and 20 Squadrons). Harriers replaced the Tornado squadrons in 1992 and stayed until 1999, when the RAF base was closed to become a civilian airport.

The family had to move house and we went from our nice house in Avenwedde to a flat on the third floor of a block of flats in Weeze, a few miles from the base. Sam and I hated it, especially with young boys and a dog. So, when we were offered an extra year on my tour in Germany, I turned it down.

In 1993, the World Trade Centre bombing happened, and a dinosaur craze started with the release of the film Jurassic Park. I don't really remember anything exciting happening apart from the usual detachments to the UK for low-flying exercises and our winter trip to Sardinia, which was a much less fun place than when I went during the summer months with the Jaguars for so many years. I was now old enough to apply to be a branch officer engineering, and after one of my annual appraisals, I was encouraged to apply by the SENGO, Vince Thomas. I really had already been through the selection process three times and failed each time, so why would this be any different? But they told me that applications for branch officer from serving personnel were treated differently and assessed under a different criterion. The Officer and Aircrew Selection Centre (OASC) had moved from RAF Biggin Hill to RAF Cranwell in Lincolnshire. Cranwell was the officer basic training centre and a flying school at that time.

So, I went again, survived the entire process again, and was told a few weeks later back in Germany that I had failed—again! The SENGO was very supportive and surprised that I had not passed, but he assured me that 'class always shines through', and to apply again in a year or so. But that was not going to happen. I have a reputation for being dogged and determined, but even I decided that after so many attempts, the officer path was just not going to work for me as there was something about me that they just didn't like.

A year later, the Channel Tunnel opened. It was the official end of apartheid in South Africa, and Amazon, some little book selling outfit, was founded. We were on detachment to the US Air Force base at Incirlik in Turkey on a mission to maintain a no-fly zone over northern Iraq, mainly to protect the Kurds living in that area. It was a strange situation as, on certain days (known as 'Down Days'), we were not allowed to fly or even be at work. So, from the golf course or the swimming pool, we would watch the Turkish Air Force Phantoms take off, fully loaded with bombs, to go and drop them on the very same Kurdish people that we were protecting the rest of the time—the madness of politics!

Then, back in Germany, I got the sad news that my grandma in Wales had died. We went back to the UK and left the boys with their grandparents in Norfolk. Then, Sam and I drove over to Wales and the funeral took place in the little village chapel.

A few years earlier, during a visit and whilst we were shopping in the local town, my grandma had taken me into a solicitors to sign some documents that she had prepared. Despite my protests, she insisted that she did not want either of her sons to have it, and her only condition was that I would sell it and buy a house for my family in Yorkshire, which was always where I had wanted to settle. So, I owned her cottage in Wales and when she died, this was always going to cause issues.

Firstly, my aunt Pam, who had owned the cottage originally, but died of cancer at just forty-two years old, had passed it to my grandma, even though she apparently had an agreement to give it to the daughter of the local farmer, who had given her such help over the years, so they were not happy. Secondly, my father and his brother also had expectations of inheriting it and were unaware that it was not in the will until grandma's death.

At the time, a radical Welsh nationalist group called Meibion Glyndwr were targeting holiday homes, especially those owned by people from England, and so it made sense to sell the property quickly, which I did. I had £40,000 in a building society, waiting for the day we could buy a house in Yorkshire.

Three months later, we were moving house back to the UK and to the most bizarre posting I would ever have – RAF Spadeadam in Cumbria. Before we left, Benson had to be taken away to spend six months in quarantine in some kennels in Cambridgeshire, and it was heart breaking to put him into a dark van as they took him away. He must have been so confused, lonely, and maybe wondering why he was being punished. He would spend the six months in a wooden kennel with just a concrete strip runway area about ten feet long that was all wire fenced in, with no walks, no proper exercise and no real human contact. We had paid a considerable amount of money to pay for this cruel experience, which changed Benson forever.

Years later, when the quarantine rules for pets coming from Europe were changed, as no case of rabies had ever crossed the channel unnoticed, I was so pleased and I wondered how many pets had died needlessly through a cruel regime that was perpetuated by the government, the likes of Crufts, and, of course, the quarantine kennels and animal transportation firms making a fortune from the system that cost pet owners £7 million a year in fees. It was finally ended in 2012, when it was accepted that rabies was not the great threat to the UK the kennel owners and transport companies were making it out to be. A science-based pet passport scheme supported with a blood test and animal inoculations was introduced. The old rules had been in place since 1897, and according to the BBC, under the new rules, it was estimated that there would be one case of rabies in a pet in the UK once every 211 years, with the possibility of a person dying from rabies obtained from a pet once in every 21,000 years.

Despite the dire warnings from quarantine kennel owners, there have been no cases in the UK since the change in 2012, other than two people bitten by animals with rabies whilst they were overseas.

11. RAF Spadeadam—An Alternative Universe

RAF Spadeadam is located a few miles off the A69 trunk road between Carlisle and Newcastle, near a village called Gilsland, close to the border with Northumbria. Built in 1955, with the original intention of being a nuclear missile silo, it was also a test centre for the Blue Streak missile project with British Oxygen Company producing the liquid oxygen on-site, missile component testing sheds, and some massive concrete engine mounting stands to test fire the engines. The abandoned remains of the missile silo excavation, concrete engine stands, launch pad facilities, and some of the test buildings are still present on the site. In 1976, the RAF took over the site and it became Europe's first Electronic Warfare Tactics Range.

Environmentally, like many military ranges and exercise areas, it is an important contribution to conservation having a peat bog, otters, all three species of British newt, and red squirrels. A mock airfield has been created with all sorts of old aircraft dotted around the place, along with dummy air defence missile batteries to give attacking pilots a sense that they are approaching a real enemy airfield. To add to this illusion, ground crew fire GTR-18 Smokey Sam lightweight unguided rockets up after the aircraft has passed over, leaving a white plume of smoke to simulate the launch of enemy surface to air missiles.

For me, RAF Spadeadam was a really weird place and when I arrived at the Mechanical Engineering Flight (MEF), I was beginning to think that I should have accepted the extension to my tour in Germany and stayed at Laarbruch. I had one SAC armourer in my charge—I won't name him to save embarrassment as he did not exactly shine on my watch. He enjoyed re-enactment days and painting toy soldiers in the small armoury where we worked. I was ok with that, providing that he had done his duties in servicing the weapons that we held first. The trouble was that he didn't do that very well and kept leaving the internal weapon stores' doors open, alarms off and he failed to wire and padlock the weapons in the gun racks. I gave him a verbal warning the first time he did it, a proper bollocking the second time. But when he left the safe open, I had no choice but to discipline him officially with a charge, the one only time I had to resort to using my rank in twenty-two years of service.

My boss was an old mechanical transport flight sergeant, who was close to retirement and planned to emigrate to New Zealand and his deputy was a Scottish chief technician general engineering guy, who turned out to be a real snake in the grass and, as I learned later, was universally hated by his own tradesman colleagues. In charge of all of this, as Officer Commanding MEF was the feeblest officer I would ever meet (and who was such a massive Star Trek fan that his wife made him the Starship Enterprise crew uniform so that he could go to Star Trek conventions dressed up as Captain Kirk), he was also into re-enactment events and went to them with my SAC armourer. So, come the day of the charge, I marched the SAC armourer with his escort into the office of OC MEF and the boss started reading out the charge, to which the SAC pleaded guilty. Then, it came to awarding the punishment and I can't even remember what it was, probably because that memory has been surpassed by the one seeing the boss getting misty eyed and fighting back the tears as he sentenced the SAC armourer. I was stood there, looking in disbelief, and thinking, *what the f*ck!*

My main job was visiting Air Training Cadet (ATC) squadrons all over Cumbria and north-west Lancashire to service their .22 rifles, which were based on the old Le Enfield .303 World War I vintage weapons. As the ATC squadrons only convened two evenings a week, this meant that I often had to stay over in Blackpool and do several locations each night. This gave me the daytime free to play golf or do my Open University studying.

When the MT flight sergeant retired and left, they did not replace him immediately and they let the GEF chief tech guy run the place. The first thing he tried to do was stop my overnight stays and make me do day trips after being in the armoury first thing before I left. He spent quite a bit of time and effort writing a policy on how I should deliver my ATC duties. I wrote a response that showed that what I was doing was already the most cost and time effective way, and his stupid and mean idea got the boot. He also hated the fact that he couldn't just walk into the locked armoury to catch me doing my OU studying, even when SAC armourer told him I was. It was not a problem to me as I would only study if I had nothing else on. So, in the end, I played the game; and if the doorbell rang, I made sure that my OU materials were hidden away and I would have some weapon servicing manual open on my desk when he came in to see me for some spurious reason. But all this just made him even more determined to get me and he had his chance when my annual performance review came around.

There are three categories covering all aspects of trade skill performance, general performance and another that covered things like dress standards and attitude, with each category having a maximum score of nine. The system had, over the years, inevitably suffered from grade creep and basically, if you did not get at least three eights, then there would be little chance of promotion. There was also a statement on promotion with 'Special Recommended' and 'Highly Recommended' at the top.

I had never been given a special recommended, but I had usually had eights, with a few nines and always highly recommended for promotion. The chief prepared my appraisal and gave me an eight and two sevens with just a bland 'Recommended' for promotion. This was a career death sentence and when I argued with him, he said that my shoe toecaps were not bulled to a shine. I pointed out that nobody bulled their shoe toecaps after basic training and that this requirement had not been explained to me when I arrived, but he didn't care. I also complained to the OC MEF, but of course, I had just charged his SAC armourer playmate and so that was that. I was friendly with the flight sergeant administration guy, and he told me that it would be treated as a 'rogue' assessment if I got my appraisal scores back up the following year and kept them there for four years. This hardly seemed fair as I had such a great track record, but that was just the way it was.

Not long after I arrived, I kept getting messages that my unsavoury uncle who had kept calling me from Spain, where he has lived, on the run from the UK law since robbing and wounding a milkman many years before. I was told that I had to return the call and that I could use the service telephone in my armoury.

When I called him, he said, "Me and your dad have been talking, and we think that you got mum to change her will when she did not know what she was doing, so this is what you are going to do...."

He then told me that when the cottage was sold, that I would transfer all of the money into my father's bank account, and that he would be given his share from that. When I pointed out that the house was not in his mother's will and had been gifted to me several years earlier, as grandma did not want her sons to have it, he got angry.

He said, "Listen, you know I hurt people and I know people over there with guns. I can send them after you if you don't do this."

Although this was quite a chilling thing to hear from my 'loving' uncle, I had my wits about me and said, "So, Peter James, you are threatening to have your own nephew killed if I don't give up the house money to my father?"

To which, he said, "I'm not threatening you, yer little shit, it's a promise."

Now, the irony of being threatened by people with guns as I sat in a room surrounded by lots of semi-automatic weapons was not lost on me, but he had walked nicely into my trap.

I said, "You do realise Peter that all calls to and from Ministry of Defence facilities are recorded and that this one could well be of interest to the UK police and Interpol?" The line went dead, and I never heard anything from him or my father on the matter again.

I was taking Benson out one afternoon for his late walk as usual and I noticed that he was limping and making a pained noise with each step. I talked with him and his tailed wagged as always, but I could tell something was not right. He often picked up a sharp stone or thorn in his foot pads and I would get it out for him and he would be fine. Sometimes, they bled and I would clean it up when we got home. But this time, he did a small yelp of pain as I lifted his back paw, and so I realised it was nothing to do with his paw. We walked on, but it was clear that he was in pain, so we went home, and I took him to the vets in Carlisle.

The vet examined him, did an x-ray and some tests, and gave me some medication for him. He called me at home a few hours later and said the test results were in and he was so sorry to tell me that Benson had some sort of cancer. He was eleven years old, so getting on, but he had always been super fit and without any fat on him. I was shocked and asked what could be done.

The vet replied, "Nothing, I'm afraid. He is riddled with it and will not last long. It's amazing he has not been suffering sooner, but he was in really good shape so maybe that helped. All we can do is make sure he is comfortable living, and if the pain gets too much, to let him go peacefully."

I thanked him, put the phone back on the cradle, fell back onto the settee, and blubbed my eyes out like a baby. I felt awful but I was also convinced that the six months in solitary confinement in that cage had broken his spirit, weakened him and the evil disease had invaded his body then.

Lying on the settee, thinking about these things and crying with anger at the system and frustration with the situation that I was helpless to do anything about, there was suddenly a wet nose on my cheek and Benson was licking the tears off my face. That kind act made me feel even worse, as I knew it must have hurt him to get up from his basket in the other room; and yet, he made the effort, realising that I was in distress over something. The meds helped Benson and the next day, he seemed more like his usual self and as we walked through Gelt Woods, I found myself hoping that the test results were wrong and that he would last much longer than feared.

But the following morning, he let out a small yelp and I knew in my heart that it was over. I called the vet and despite it being a Sunday morning, he would open his surgery that afternoon. It felt like being on Death Row for the rest of the day. Sam kissed and hugged Benson, trying not to cry before I took him, but we decided not to tell the boys yet, who were thankfully oblivious to his condition. I drove as carefully as I could to the vets so that he didn't move much in the passenger floor. I glanced at him and rather than being nestled down and asleep within ten minutes, as he usually did, he had his head on the edge of the seat and was looking at me. I couldn't look at him again. I felt like I was murdering one of my best friends.

We arrived and the vet gave Benson some fuss. Normally, he hated going to the vets and always looked to me for reassurance, especially when I always had to lift his substantial weighty body onto the vet's examination table, but this time was different. He was calm, almost as if he knew the vet was trying to help him.

The vet asked me if I was ready and, "Did I want to say goodbye?"

I took Benson's head in my hands and kissed his snout and, looking into those kind brown eyes, I said, "I'm so sorry."

I saw a shadow pass over his eyes and then, in my head, I just heard a voice I had never heard before that said, "it's okay." It was probably the vet saying it, but it seemed to come from Benson.

And then, just like that, and far quicker than I expected, the vet said, "He's gone, and you did the right thing by not being selfish and making him suffer."

I tried to get my wallet out to pay the bill, but the vet would not have it. He also offered to take care of the body, but I felt Benson deserved a better resting place than that. So, I carried him out in a blanket and carefully placed him in the back of the car and drove home with tears streaming down my face, but at least I knew exactly where I was going to bury my canine mate. I called in at home and got a spade and let Sam know what had happened and where I was going. The boys were playing out, so at least I didn't have to deal with them yet.

I then drove the twelve miles of my daily commute to RAF Spadeadam and on the way, I revisited all of the memories with my dog—rough shooting; him alerting me to ducks that I couldn't see flying towards us in the dusk sky; playing football with him and the boys; or when he destroyed the furniture as a bored puppy; or when he crapped on Sam's favourite sheepskin rug when he was not well—the memories all came flooding back.

I arrived at the camp and up from the administration area of the camp is an operations centre called Berry Hill, and it is way up in the hills. Halfway up there is a junction and sited there is a small explosives storage building that I visited frequently. The view from that point is magnificent—looking over towards Hadrian's Wall and across the river Irthing valley, and with Pennines on the horizon. I thought that the place was perfect for my boy. I dug a hole and was thankful it was Sunday and that nobody was driving by. I carefully buried him and said a prayer, even though I'm not religious—it was just in case he needed one to get into doggy heaven. I did not mark the grave in case burying pets on MOD land was a problem and I didn't tell anybody. That night, we told the boys, who were obviously very upset, and I had a really bad dream in which I had not buried Benson deep enough, and foxes had dug him up to feast on him.

The next morning, I left for work early and drove straight up to the explosive storehouse to check the grave and was relieved to see everything was fine. I was only at Spadeadam a few more months, but every time I visited the explosive storehouse, I said hello to my boy and I like to think he is happy with my choice for his eternal resting place.

Years later, I was driving through the area on the A69 Carlisle to Newcastle trunk road with Jules and, as I saw the signs for RAF Spadeadam, which might have well said "Benson's Burial Ground," I started to cry quietly to myself. She asked me what was wrong, and I said, "My dog Benson is buried over there." Not being an animal person, she was shocked how emotional I was over the loss of a dog over twenty years earlier. But then she understood that for me to get like that, Benson must have been one hell of a dog—and he was. I have had several dogs in my life and loved them all, but there was only one Benson. Writing this book has stirred many emotions in me, but this part was particularly hard for me to write, and I will admit there were a few tears doing it! It would be eighteen months before Sam and I decided that we were ready to have another dog, and I still remember feeling disloyal as we chose from a litter of black Labradors a new puppy that we called Jake.

One of the benefits of having my non-job at RAF Spadeadam was that I had plenty of time on my hands. Along with doing my OU studies, I also organised golf expedition trips to Scotland for the golf-playing officers and lads of the base, and we had a couple of great trips to the St Andrews area, playing several courses over a long weekend. In addition, I volunteered to set up and run the combined mess social committee and we planned and delivered an excellent all ranks Christmas Ball.

1995 arrived and heralded the end of the Bosnian War and the film 'Toy Story' was released. When it came time for my annual appraisal, there was a new OC MEF in place who was the opposite of the Trekky previous boss. The new boss was Flt Lt Ted Atkins, who was a legend in the mountain climbing world and a great guy. We really hit it off as he liked me and didn't really like the GEF chief tech. So, when the chief complained to him that I was spending my time in the armoury doing my OU studying and gave me crap appraisal scores yet again, I was called into Ted's office, and he told me that he had used his discretion to alter my appraisal scores up to three eights and a highly recommended. He also told me that he had also used that discretion to downgrade the chief's appraisal scores to sevens, as for the past two years, he had been writing his own and giving himself nines and special recommendations that the previous trekky OC MEF had just signed off—that was a great schadenfreude moment for me!

He then said, "Sergeant James, I am ordering you to complete your Open University studies during working hours, in the armoury if necessary, providing that you are satisfied all other duties have been completed."

This was great news and when I left his office, I had a beaming smile on my face that the chief sat in his office across the corridor saw and made him confused. Ted Atkins also quickly got a replacement MT flight sergeant in post and so the chief tech was effectively demoted back to just running the GEF bay—result! A few years ago, sadly, Ted was killed by falling from a mountain in France and it was newsworthy enough to make an entire page coverage of his life in the Times newspaper obituary section.

In 1996, the year that would see "Dolly the sheep" cloned, I was packing my bags to go to USAF base Incirlik in Turkey again, but this time for six months to run the armament support section as part of the operation to enforce the Iraq no-fly zone. It was so good to get amongst regular RAF people, real aircraft (the Harriers had now been replaced by Tornado aircraft) and be handling real bombs and weapons again. I loved every minute of it, especially as we were normally done by 3:00pm and so could go play sports, such as tennis, football and golf. I was also using the trip to lose weight and get fit, and I came back eighteen pounds lighter than when I went.

Unfortunately, I was sent back early after four and a half months as I broke my wrist in two places playing football. I had served long enough that it counted as my long detachment, which should have been in the Falklands until my admin buddy found me the role at the much more amenable Turkey. My role at RAF Spadeadam had been so vital that they didn't even get a replacement in for me whilst I was away, just getting the chief authority to supervise the SAC armourer as he did my duties. When I came back to the alternative universe of Spadeadam, nothing had really changed, and I slipped back into the dreary routine at work—driving up and down the M6, visiting ATC squadrons again.

After the success of the Christmas Ball, we organised an Open Day BBQ to invite the locals, put on a BBQ, a local band and free beer for a ticket price of £10. It was a disaster of a day as our people failed to control the beer and the local lads were just filling their car boots with slabs of beer. There were also drunken louts fighting and some inappropriate sexual behaviour with the local girls. The next day, the MT hanger that had staged the event looked like a bomb had hit it and it would need to be cleaned up before the working day started; but none of the committee clean-up party had turned up.

Fortunately, there were about twenty army lads staging through and sleeping in the hangar, so I offered them twenty slabs of beer if they cleaned it all up. Two hours later, all the rubbish was in twenty full bin bags and the hangar floor had been swept. But that did not change the fact that the event had got out of control, and I took ownership for that and offered my resignation from the social committee. The GEF chief tech, who had previously shunned being on the committee and didn't even attend any of the functions, now stepped forward to take the chairmanship of the social committee and made sure everyone knew that he would have run the event very differently. The chief got on very well with the new MT flight sergeant, who unfortunately remembered me from my Marham days, and how I had wriggled out of a charge when I damaged a service vehicle. So, when my annual appraisal came around, I was once again facing lowered scores with the disastrous open day well-documented in the form.

Fortunately for me, one of my golf trip guys was the squadron leader for operations at RAF Spadeadam, and unrequested he had written to Ted Atkins to tell him how well both golf trips had been organised, but the real saving grace came when my extremely complimentary Incirlik in Turkey detachment performance report came through, which the flight sergeant tried to ignore and defer to the following year's review period. But Ted Atkins would have none of it and so my scores were back up to straight eights and highly recommended again. But I knew that this could just not carry on and as my flight sergeant admin mate was now at PMC in RAF Innsworth.

I called him and asked for a favour, "Please get me a posting to anywhere front-line away from here."

An hour later, he called me back, "Not front-line, but close. The Weapon Training Cell at RAF Leeming has a post becoming vacant soon and your qualifications make you a perfect fit, do you want that?"

I thanked him profusely and said, "Yes, please."

Who needs a 'Dream Sheet' when you have a mate in the right place? The MT flight sergeant, no doubt, influenced by the chief, tried to block my posting as my replacement would not be in until three months after I had gone, but when it was pointed out that they had lived without me for longer than that when I was at Incirlik, he couldn't argue. In January 1997, I was elated that we were driving south and back into the real world. I heard that after I left Spadeadam, someone had reported the chief tech to the special branch of the RAF Police for using a service vehicle to drive to Nottingham twice to collect and return a personal computer. Apparently, they even flew all the way down to the Falkland Islands, where he was on detachment, to interview him. I don't know if he was charged, but I did notice that he didn't get the promotion, which he was so desperate for, and was last seen working for Network Rail in Lancashire servicing electrical plant on the railways.

12. Yorkshire at Last

I was so happy to be heading to RAF Leeming and be back at a real RAF base. RAF Leeming is on the A1 trunk road between London and Edinburgh and was built in the late 1930s as part of the build-up to World War II. During the war, it was a bomber station for both the RAF and Royal Canadian Airforce before becoming a night-fighter base, until, in 1961, it became a flying training base. In 1987, HAS dispersals were built to house three squadrons of Tornado F3 aircraft—11 (F), 23 and 25 (F) squadrons in a QRA role for the next twenty years. In 2008, the base reverted to becoming mainly a training facility with an additional role of a deployable expeditionary air wing. In 2020, the Yorkshire University Air Squadron relocated to RAF Leeming, when RAF Linton-On-Ouse was closed.

Before we arrived, we had bought a nice house on the outskirts of Ripon that had a huge garden but needed updating inside. Sam had used her estate agent skills to negotiate a considerable reduction in the price and had been helped by the old lady that owned the house, who love the two boys and Jake, the young Labrador, during our inspection visit. She accepted our bid and even turned down a higher one as she liked the idea of our family living in 'her house'.

Tosh, as always, rocked up to help me fit a new kitchen and we replaced both the main bathroom and main bedroom ensuite bathroom. We were a good pairing and called ourselves 'bodge-it and scarper', which is probably why, at regular intervals, Sam and Rosie would come and 'inspect' our work and point out all of the faults that had to be fixed before we could carry on. Never before has a spirit level been used so extensively to settle arguments.

I still get this now with Jules, and I have even put up a shelf that was according to the spirit level, perfectly positioned. Jules made me tilt it slightly at one end because 'it doesn't look straight when its level!' This house refurbishments had just been finished when I was sent away on training courses, the first to the manufacturer's plant to learn about the new ASRAAM missile, then a boring course on manual handling techniques from which the only useful learning point was about how toddlers have perfect lifting techniques when they totter about and then pick up something from the floor because they keep a straight back and bend their knees to lower themselves to pick up an object rather than just bend over, which stresses the back muscles!

Then I was off to RAF Cottesmore for a month to do a Tornado F3 Weapons course before, finally, I could start work in the Weapons Training Cell, mainly teaching new squadron guys how to do Tornado OTRs. The guys I was working with in the WTC were Mark Godwin, who I regarded as Mister EOD and Steve Aldred. The armourers I can remember from across Leeming include Andy Sattherthwaite, John Rowson, Brett Logan, Tony Boardman, Mark Kaye, Brian Godfrey, Steve Greenough, Paul Whitbread, Dave Lowe, Graham Tate, Ron Thompson, Rupert Webster, Dave Gibb, Al Haresnape, John Williams, Mark Bean, Mick Haygarth, Phil Lowther and Mick Fisher.

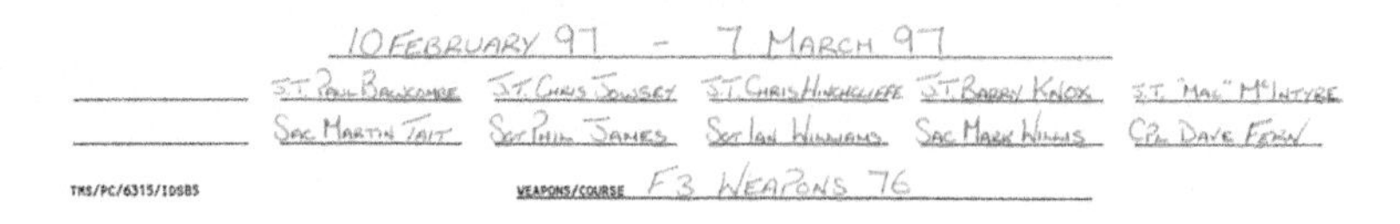

Fig 34. The guys on my Tornado F3 Weapons Course
(I am 2nd from left front row)

With hindsight, 1997 turned out to be a momentous year with Tony Blair becoming UK prime minister in May, after leading the labour party out from the political wilderness. He then had to oversee the handing back of Hong Kong to the People's Republic of China in July, before having to lead the nation in mourning after the loss of Diana, Princess of Wales, after a car crash in Paris at the end of August.

On a personal note, November that year also marked the culmination of six years' study at the Open University for me, and I now had a 2.1 BSc Honours degree, mainly in systems management, design and innovation to show for it. As well as attending the degree award ceremony with my family, Harrogate Town Hall also hosted an afternoon tea event for all the OU students in their council region that had gained their degree that year. So, I found myself sipping tea with about thirty other guests, when an elderly man approached me and asked which branch of the services was I in. This surprised me as I was not in uniform, and when I told him RAF and asked him how he knew, he told me it was my bearing that showed I was a military guy. He also said that now the RAF would soon make me an officer. I explained that I doubted that, as I had tried several times and never been successful.

But he said, "Ah, yes. But now, you are an educated man, dear boy. It will happen."

I doubted that very much, but it turned out that he was right; by coincidence, the OASC had put on a roadshow at RAF Leeming, which my new boss OC Weapons Engineering Flight, Flt Lt Heath, had attended. He explained that he now had a sergeant that had better qualifications than he did. Late afternoon that Friday, I was called in to see the boss; he told me that as I was now apparently an embarrassment to the RAF recruitment system—all I had to do was apply for a branch officer engineering commission, and after a tick box trip to OASC at RAF Cranwell, I would be promoted. He told me that I needed to let him know on Monday so they could arrange things, so I should go away over the weekend and think about it. You would think that I would have been delighted at this news; after all, it was what I had wanted for the past twenty years. But the more I thought about it, the angrier I got!

I had not been good enough over the past three attempts to get a commission, but, now, despite being exactly the same person, because I had a piece of paper saying I had a degree, suddenly it was a done deal! On the Monday, I went to see OC WEF and handed in my general application form, but it was not an application for a commission; instead, I was applying to leave the RAF seven years early by giving in my eighteen months' notice. The boss tried to talk me out of it but my mind was made up—I was leaving the service I had loved being a part of for twenty years.

To be honest, as soon as I knew I had passed my degree a month earlier, I had been thinking about leaving. I think the actual moment I made my mind up was on detachment with one of the Tornado squadrons to Florennes Air base, near Charleroi in Belgium.

We were in a bar, singing RAF songs, and one in particular is called the 'Shackleton Song', which ostensibly is about an aging four propeller engine plane. But it is designed to get young new additions to the squadron to be encouraged to pretend that they are one of the engines, and stand there like an idiot swinging their arms around to simulate a propeller. Then, when one or more of the hapless volunteers are fully engaged, swinging their arms around, there will be a warning shout that there is a fire in whatever engines the young – and they are always young volunteers are simulating. This will result in the 'flames' being doused by everyone's pint of beer, which, of course, soaks the unwitting participants. It is always funny to see their shocked faces; initially, at getting soaked and then the realisation that they have been had. How they react to this is important as anyone that gets angry and can't take the joke is probably not really squadron material. I must have witnessed this initiation ceremony at least thirty times during my career, but this time I thought to myself, *James, you are thirty-nine years old; you have a degree. You could get your RAF pension next year if you leave the RAF. Perhaps it's time to grow up, put your big boy pants on and go out into the big wide world and get a real job!*

A few weeks later, after the certification OTR load of another team, I was walking with the OC Engineering Wing, Paul Raine, who was along to witness and certify the load. He asked me what my plans were, and I told him that after getting my degree, I was thinking of leaving the RAF.

To my surprise, he said, "Good, you are wasting your time doing this and can probably make far more of yourself in civvy street."

This really surprised me but also encouraged me at the same time and so the offer of the 'done deal' commission was actually the insult that made me decide to leave. As soon as my eighteen months' notice went in, everything changed. My focus became all about resettlement training and preparation for a new life.

With 1998, came the Good Friday Agreement and its terms, some of which meant forgiving past atrocities on both sides, which was not easy for everyone. The company Google was also founded, and in a few years, it would change the way we all search for information. Meanwhile, I entered the forces' resettlement system. I thought that the resettlement training provided by the RAF at the time was pretty good and the emphasis on transferring our service skills to the civilian working environment was very valuable, as it gave me confidence that I was not just starting out but adding my service skills as a contribution to my new employer.

The RAF also kindly bent their resettlement rules to let me take all eight of the health and safety and also their environmental management courses that ran at RAF Halton, which would later really help with me getting an NVQ degree equivalent. It turned out that Vince Thomas, my old SENGO on 3 (F) Squadron, was now the group captain in charge of the RAF Training School and I suspect that his permission was needed. I had realised that unless I went to Saudi or UAE to work in their air forces, or did landmine removal work, that I needed a completely new career. There was not much demand in civilian life to put rockets and bombs on cars and I had only ever seen a certain Aston Martin DB5 fitted with an ejection seat. I had already chosen a new path by getting into health and safety and environment, which I had seen grow in importance in the RAF over the past decade and started getting qualifications in the field.

After two years of distance learning, I had my NEBOSH Diploma Part 1. Through the NVQ route, I would have Part 2 a few years later and become a Chartered IOSH practitioner, which opened doors to decent health and safety roles. At the time, it seemed to be a natural fit to being an armourer, in which, dealing with explosives has a safety mentality built in.

It also helped me being settled in Ripon in North Yorkshire for a couple of years before my discharge. I now had contacts and civilian friends that also supported me in what is a dramatic life change – moving from a supportive forces environment, in which many things are decided for you (such as where you will be living, job type and for how long) to civilian life, in which you really are on your own and your employment is at the whim of employers who are only focussed on one thing – profit.

In life, you need luck and I have had plenty of it; fortunately, with most of it good, except for property investments. An example of my good luck was meeting Doug Wentworth, who, despite coming from Liverpool and an ardent 'red', has become one of my best friends. We met watching our kids learning how to ride horses and started chatting. He told me he was in the timber trade; at first, he didn't tell me he owned a very successful timber mill! I told him that I was thinking about leaving the RAF and was unsure if I should try and set up my own health and safety consultancy or get a job in a company. His advice was to get a job in a consultancy company and learn how the business worked and then go out on my own. He didn't realise then that he was to become my business mentor, who helped me get pay rises and promotions by using clever negotiating strategies, as well as sound business advice. We became good friends.

Unlike me, Doug is one of these really sociable people that naturally attracts others, and so he has many friends. But I am content to know that I am an important one to him. I think our friendship really firmed up when we both went through our divorces and saw that people whom we had previously considered to be our friends, step back. But we both stepped up for each other, and now, despite me living away for twelve years in the Middle East, we maintained our close friendship, and I'm sure we always will.

On that first day at the riding school, Doug told me about a group of lads that he played five-a-side football with every Thursday evening and that was how I was introduced to the Ripon Round Table guys. I soon joined Ripon Round Table, which is like being in the Rotary Club but with far less money, less pomp and a much greater emphasis on drinking alcohol and having a good time, as well as raising money for good causes. It met every other Tuesday, in a pub naturally, unless there was an activity. So, every other week, I would find myself doing something unusual, such as ice skating in Bradford followed by a curry; white water rafting in the Tees barrier; going to comedy nights; clay pigeon shooting; canal boat cruises; or anything else that was dreamt up. This, together with occasional weekends away with the Round Table guys, regular parties, BBQs and playing golf at Ripon was the bedrock of my social life.

Every Christmas was our main fund-raising push for local worthy causes. We would get out Santa's sleigh and go around Ripon, handing out sweets to kids, who could also get on the sleigh to have a chat with Santa about their Wishlist, as we were collecting money from their parents and knocking on every house as we went. We had to carefully plan this in coordination with the Ripon Lions Club, who did the same thing.

One year, there were two Santa's sleighs in the same street, which must have been a bit confusing for the kids! One night, it was howling wind and rain and the four-man team decided that rather than go out, they would rather all just put £50 in the kitty and go to the pub – *sorry kids, Santa is taking a break at The Royal Oak*. We also set up a rather dodgy wooden box Santa's Grotto on the market square each year and had Santa sit inside to meet the kids, whilst we fleeced their parents outside for the dubious privilege.

Twice a year, we would all attend Ripon races with buckets to raise money from the punters who used to come down from the northeast in buses in droves. Once, after several hours, we had only raised about £100, which was disappointing. So, one of the guys had the great idea of asking his teenage daughters to dress provocatively and come down in time for the last race. The girls came wearing some rather revealing clothing and stood on either side of the exit gate to the car and coach park. The drunken Geordies obviously really appreciated their efforts as we had another £500 within an hour! I think it's fair to say that the Ripon branch of the Round Table was a bit non-conformist and not really a participating branch of the organisation, as our focus was on having a good time.

We rarely participated in the wider Round Table events, such as annual dinners. I remember an official visiting one of our meetings to find out how we were so successful at having a robust membership and were pretty good at raising money for our chosen good causes. The regional official was clearly disappointed by us because, at the end of the meeting, he told us off and said that he was not very impressed that we did not religiously follow the standard meeting script, or read out the oath. And, he was appalled that we had decided to sponsor a schoolgirl going on a trek in the French Pyrenees to raise money for the disabled, providing that she did not come and tell us about it afterwards (to be fair, we were only half joking about that). At forty-one years old, you must leave the Round Table and so most of us left around the same time and formed a Ripon 41 Club that was much less formal and only met monthly. It kept going for several years, but at a much more leisurely pace; whereas I think the Ripon Round Table group folded not long after so many of us left.

In 1999, as the end of a tumultuous century approached with fears of a millennium bug that would disrupt the world growing, but the new president of Russia, Vladamir Putin would later go on to do just that by invading Ukraine. By May, I had completed all of my resettlement training and on 13th June 1999, I left the RAF after twenty-two years and six months of service to start a new chapter of my life.

13. Life After the RAF

I wasn't sure if I should write anything about my life after leaving the RAF. But then I decided that it was relevant as the RAF shaped the man I had become; it gave me opportunities in my future civilian life that I took full advantage of and it also provides a comparison of the two different working lives that I have had. However, if all you are interested in is my RAF days, then please feel free to stop reading here.

When one of the guys at Round Table, called Simon McCudden, heard I was leaving the RAF, he suggested that I come into Leeds to meet his boss, who was a partner called David Rigby at EC Harris. It was a consultancy mainly in construction and facilities management. I found out as much as I could about the company and prepared well for what was my first interview in twenty-three years since joining the RAF. It went well, and by the time I got home, Sam said that they had phoned and offered me a job. This was an enormous relief to us both. By choosing to leave the RAF at forty, I had given up on the idea of becoming an engineering officer, and even if I had stayed as an armourer, I had seven years of guaranteed employment; and if I had been promoted twice, I could have stayed in until I was aged fifty-five. I also felt justified in the six years' effort I had put into getting my BSc degree through the OU, which the RAF had kindly paid for, as this opportunity would not have been open to me without it.

After over twenty-two years of having my life pretty much decided for me, now I was out in the big wide world and desperately having to adjust to how a civilian company worked. EC Harris was a very well-run company, but without the sort of processes that are required by a quality management system, which, of course, has its roots in the military as a result of the need to standardise things like ammunition. So, it took me a while to get used to the absence of process, checking and supervision oversight on my day-to-day work, but I soon realised that this was also liberating.

On my first day at EC Harris, they gave me a little Vauxhall Astra company car and I was put in a team of four H&S planning supervisors that had a mainly administrative role on construction projects—something that would change for the better a few years later. I was put on a project supporting the Asda supermarket chain and found myself driving all over the country to carry out H&S inspections on their new build store projects and renovations of existing stores. The audit template that I had was not very good and so I developed a far more comprehensive one and started to produce reports with statistical analysis, which the client appreciated. I was also sent down to a woollen mill in Dewsbury, who were being charged by the Health and Safety Executive for unsafe practices. I ended up doing several weeks of work developing H&S manuals, policies and new working practices for them—my first HSE consultancy job.

I have already explained why I chose a second career in health and safety, but my god, after the excitement of the being in the forces and working with competent people, it was hard at first. I have met some of the most uninspiring people in the HSE field, who were great at quoting rules and regulations to stop you doing something, but were unable to come up with a solution to get around the problem. My approach was to try and think of the solution first and then justify why it was needed because of the HSE requirements. This approach was appreciated by clients and although my solution would often have issues, it set the client thinking on the right path. As a result of my style, clients liked me, and I got lots of new work.

A year later, I was leading the team in Leeds and two years after that, I was made an 'Associate', which was like being a SNCO in the RAF. I was also promoted and given a brand-new VW Passat as my company car. I had a very diverse work life as I was appointed HSE construction advisor on the Feltham Town Centre regeneration project; The Queen of Speed ride at Alton Towers theme park; Lead Construction HSE Auditor for Yorkshire Water on their AMP3 programme; and advising Tate & Lyle on their facilities management HSE risks. I also developed the HSE policies and procedures for Her Majesty's Court Services following their restructuring, along with those for The Royal Courts of Justice. I also had one-off tasks, such as a fire survey for Birmingham City Hall, which was a fascinating Victorian building.

EC Harris had been formed as a family firm in 1914 and had grown into a global company, but it did not have any proper HSE systems in place or even a decent policy, and they didn't really care as they just saw it as a distraction. I wrote a memo to the Chief Operating Officer, Phillip Youell, stating that the board members were at risk and that I was happy to develop what was needed, and that it could also help us create a revenue stream. My memo came back to me with one word written in red on it—Noted—which, I soon learned, is a corporate way to tell you to wind your neck in. But something must have changed at board level as I was then told I was being transferred to the London HQ office and would be Global Head of Safety, Health and Environment, as I had suggested in my memo.

I did not have to move from Ripon as the company paid for my train tickets to London from York and my hotel accommodation in London, when I needed to stay over. I was again very lucky in having a senior partner, Peter Madden, as my boss and he guided me on how to become partner material. Soon, I was wearing double-cuff shirts, decent suits and shoes as part of the 'act as if' approach. Peter was an avid Chelsea fan and I took him with my boys to Elland Road to watch Leeds play Chelsea. He got me a ticket for the return fixture at Stamford Bridge and I changed into my cold weather 'footy parka', but he just turned up in his suit under a trench coat, which I had changed out of! He explained that in London, they were quite cosmopolitan, and he was right – I was the only one wearing a parka. To make matters worse, Leeds lost the game in the final minute!

When I was staying in London each week for three days, I would work really long hours, go for dinner somewhere, then go back to the hotel and work some more as I felt I had to justify the cost of staying over. Each morning, I would run from the hotel near Kings Cross through the train station to my gym in Bloomsbury and be back in time for breakfast at 7:00am, dodging the early morning commuters on the way.

Within a year of doing this life, I had developed all the HSE systems needed and implemented them at all twenty-two of our UK offices, with local SHE Coordinators trained to do the day-to-day stuff. I also developed a product called SHEQual that helped smaller companies prove that they were HSE compliant, and I had got some major insurance companies interested, along with the Health and Safety Executive (HSE). But despite getting 5,000 small companies on the system as clients, a different system called CHAS, which was local government-owned and ran out of Merton, became the market leader and our product could not compete.

The company was then prosecuted over a bad asbestos incident at a primary school and I was asked to mediate with the HSE who were prosecuting us. Luckily, I had a good relationship with some senior figures at the HSE and I had also been appointed to a government CONIAC Occupational Health Committee and was on the Construction Client's Group as a HSE advisor. The resultant HSE investigation audit of our company showed that we had put in place good HSE systems recently. We were advised to plead guilty, which the board were loathed to do, as it would have to be declared in every project bid for five years, but in the end, they agreed and we only got a £5,000 fine and suspended sentence.

My time in London meant that I was in daily contact with all the board members and six years after I had joined EC Harris in Leeds, I was made a partner with a clear path to becoming an equity partner. So now, I was like an officer – something I had struggled to achieve in the RAF. One of the highlights of this was not just the massive pay rise and bonuses that I got, but that I could choose a new company car that came with a fuel card that paid eighty percent of my fuel costs. The list included BMW and Mercedes cars but then I saw that I could have a Mazda RX8 sports car, and so that was that. I was getting my mid-life crisis or male menopause issues treated by my work. It was probably my favourite car of all time, but you could watch the fuel gauge drop as you drove along, and after a blow-out on my first trip, it cost over £200 to change one tyre, so I knew that I could never afford to own one of these cars!

EC Harris then decided that, as a company, it would now actively recruit ex-forces staff with or without a degree. So, I managed to get a couple of armourers employed by EC Harris with Ken Amos running SHEQual in Leeds until it was abandoned and I sadly had to let him go, which was a dark day for me personally. I also introduced Jez Norris to the London logistics team and he impressed at his interview and got a job with them, until the commute was too much, and he went off to work elsewhere. I even set up an interview for one of my old OC Engineering officers, Paul Raines, and he became head of IT for EC Harris for a few years.

I was now leading a UK team of thirty SHE Consultants in seven offices across the UK and I implemented our systems at all our overseas offices and tried to get HSE business streams going in them too. So, I was often in Dubai, Hong Kong, Prague, Paris and our other 17 offices implementing corporate HSE but also trying to develop new HSE income streams through introductions to the clients of whichever office I was at. This global travelling life of business class flights and fancy hotels sounds exotic, and, at one level it is, but really, it is just the company's way of owning your soul and I soon tired of the travel and shallow life of visiting really interesting places but not actually seeing anywhere other than office buildings of clients or EC Harris.

Eventually, though, an exciting lead came in from Dubai to provide UK-style HSE support on the Dubai metro project. I sent out a young guy called Aziz Zerban and I briefed him that if he could get enough work there that he would lead the team. Being a talented, enthusiastic, ambitious and incredibly hardworking individual, he did just that, and so now we had a small SHE team in the Middle East. Gradually, the work built up and our HSE reputation was spreading. Through Aziz in Dubai, we were invited to bid on a project for the Abu Dhabi government to develop and run a pilot scheme and to understand what they needed to have a HSE branch for the construction sector, which, at that time, was a virtually unregulated and a deadly place to work. I led the bid team, and we won the project, which was quite something, provided I was prepared to personally lead the team and be based in the UAE. The EC Harris board agreed and thought it would be a temporary assignment, but I wanted to stay out there as the tax-free salary was massive.

The project started well but then we had problems. I was new to the Middle East and had no clue about the culture and how to deal with Arabs. It became a disaster when I lost my temper in front of the client with two Egyptian professors, who, despite knowing nothing about construction, kept giving us problems and making ridiculous demands in front of the client. As a westerner, losing your temper with anyone in the Middle East is a big no, no. As a result, I was taken off the project after six months and I realised that I had to improve my project management and cultural awareness of the region. It was my first black mark since joining EC Harris and they then told me I had to go back to London or I was risking my job.

But I did not want to go back. I found out later that my timing was awful, as the company was looking to get itself bought out and the senior partners wanted to slim down the partnership numbers so that the remaining senior partners would get a larger slice of the sale pie. In the end, around one hundred and forty partners were fired and left the business, and I was sent an appointment with the Chairman for just before Christmas, so I knew what was coming. I flew back to London and attended the meeting to get the 'hard word'. I must have surprised him as I didn't react angrily or get emotional, I simply thanked him for the ten years I had been there and the opportunities and knowledge I had gained.

To that, he said, "Phil James, you are one hell of a guy. Thank you for everything you have done, the company is in a better place thanks to you."

It seemed strange to be getting complimented and fired at the same time! At least the financial settlement of three months' tax-free pay was quite generous. What made the situation even more bizarre was that I had asked Sam for a divorce when she visited the UAE two months earlier, as neither she nor the boys wanted to move to the Middle East. To be honest, it just gave me an excuse that I, after six months of living apart, realised I was looking to end a twenty-six-year marriage. Our marriage for me had become lifeless, in which, we were more like brother and sister than man and wife. There was absolutely no blame attached to Sam with regards to the marriage failing, she was and is a great mother. It was me that had changed so much since leaving the RAF, and I was frustrated and unhappy with the relationship. Maybe if we had tried harder and communicated better, things could have been resolved.

As I was expecting to be fired, before I left for London, I told my friends in Al Ain where I was working and living, about the situation, and as good friends do, they were there for me. In particular, for me it was Ian Shaw, who was an ex-army physical training instructor working for the UAE military. I first met Ian in a golf buggy at Al Ain Golf Club, when we were randomly paired together.

When we set off, Ian asked in his broad Welsh accent, "So, where are you from then, Phil?" I told him Ripon in North Yorkshire and he said, "Really, I'm from Ripon too!"

I replied, "Not with that accent, you aren't, boyo."

He explained that he settled there in his army days and started naming people that we both knew. I couldn't believe it. What were the chances of meeting someone from Ripon out here in Al Ain, which is the UAE's inland town that nobody has heard of! It turned out that there was another ex-army guy, also called Phil, from Ripon as well, and so to avoid confusion they called me 'PFR2'.

Just before Christmas, Ian put me in touch with a small company in Abu Dhabi called Dascam, which was owned and run by a small bunch of UK army and navy ex-officers. I had an interview and was offered a job starting on 3rd January 2009, provided I was willing to take a serious pay cut. So, in effect, I was never actually unemployed. I can remember sitting on the flight to London and trying to understand why I was fighting so hard to stay out in the UAE, when everything happening to me was suggesting that I should go home. But I have always used the same process for big personal decisions, which was to use my brain to weigh up the pros and cons, let the feelings in my heart influence things but then let my gut decide, and my gut was telling me to stay in the UAE. Even though I trusted the process, I still had my doubts. Yes, I had a new job, but it would be working with Arabs that I had just messed up dealing with, and for far less pay. I was also aware that I was about to get financially wiped out in a divorce.

But deep down, I knew why I was really staying. I had met Jules a month earlier when I hosted a 'becoming a bachelor again' BBQ at my villa, and Ian's wife Sharon had invited Jules to attend. My neighbour in the same compound used the same driveway. She was a miserable woman work colleague that I regretted employing for the pilot HSE project, and when one of my visitors parked a car in her parking place, she returned and had to park in the other slot of her double car port. Not used to parking there, she dented her car and then blamed me, so I knew I had to stop any more cars from parking there. Then, Jules rocked up, a robo-chick with a mass of long blonde hair, in her sleek silver US-made Seebring coupe.

I could see that she was going to park in the neighbour's half of the driveway and that would just cause issues, so I said, "Hi, you can't park there. Can you park outside, please?"

Apparently, Jules thought I was being a bit rude and nearly drove straight home! Thankfully, she didn't, and we met up for a drink and once again I tried her patience as she turned up glammed up to the nines only to find me in a T-shirt, shorts and flip-flops. Somehow, I survived that sartorial misjudgement and we went out a few times, but she was playing the relationship thing with a very straight bat. I was rather confused by her approach, and I sought advice from Grace Doswell, my very streetwise PA at EC Harris, who told me to just keep going and that she would eventually fall for me.

Jules is a one-off and is incredibly intelligent. She is a Doctor in English, has three degrees, and despite being in her forties, Jules had never married and had no intention of doing so until she found the right guy, even though she knew it meant not having children. She had worked overseas in Germany, Spain, Mexico, as well as the UAE and I was totally smitten with her and convinced that she was why my gut had made me feel that I had to stay in the UAE.

I spent Christmas in Ripon, dealing with solicitors about the divorce and sorting out a new bank account. I was staying at my mate Doug's flat, and I had a lovely Christmas day with Sharon and Ian and their family, who were home in Ripon for the holiday too, but I felt like an intruder in their family happiness and knew that my own boys were celebrating Christmas with Sam down with her family in Norfolk. However, I kept my promise to call Jules on Christmas day, and that must have got me some smarty points as on 27th December, she sent me an application form to be her boyfriend. To her, this was a silly little jolly jape, but I treated it like a job interview and comprehensively completed it and duly emailed it back to her the next day from Manchester Airport, just before I boarded the flight back to the UAE. On New Year's Eve, we were both attending a function at the Golf Club on different tables.

Later, we got together and ended up in a nightclub, and it started properly there. I think at first, Jules couldn't work out if I was a deranged stalker or just a determined guy. In the end, she decided that as I was so utterly convinced that we were meant for each other, she would go with it. In April 2010, I married the most intelligent and beautiful woman I had ever met, and fifteen amazing years later, despite her being a Manchester United fan, all I can say is that my gut was so right!

I thought EC Harris lacked systems after the RAF, but the job with Dascam felt really strange after the sophistication and systems of EC Harris. I was now one of just ten people working out of a villa in Abu Dhabi. The company was owned by an ex-army officer who had serious pedigree in the MOD and had set up the company as a sort of people resource, training and advisory consultancy offering. The leadership were all ex-forces officers from the British army and Navy with some ex-SNCO's working under them.

My boss was David Parry, who was an ex-submarine captain and despite being very old school, was a wonderful guy, but it was like having an English teacher as your boss. I was used to the crisp, straight-to-the-point proposals of EC Harris, but Dascam liked to produce these flowery picture-book proposals for projects that took me a while to get used to writing. David would review them and write 'see me' on the bottom. I would then get the same lecture about "Who, Where, When, Why and How?" as well as the need to "edit, edit and edit." But when I did this with an EC Harris style proposal, he didn't like it!

Dascam did have a few impressive clients and the technological solutions that they had were even used on the inaugural Abu Dhabi Grand Prix in 2009. I developed the corporate HSE stuff and some HSE products, but it was hard work winning work from such a small company.

Later, the owners sold the company to a defence giant called Ultra Electronics and then we were partnered with a local Arab company. But it was not really working for me. Dascam had a sister company called Steelhenge, who specialised in the UK at providing business continuity and crisis management consultancy, and after meeting with the MD, Dominic Cockram, in London, he offered me the post of Regional Director with a remit to start up a branch of Steelhenge in Abu Dhabi.

I had developed an interest in this topic after the 7th July bombings in London, when I had been in London that day at a conference in Knightsbridge. There were no communications as mobile phones did not work and I had to walk right across the city to get back to our office. I had been classed as 'missing' and got loads of hugs from the relieved ladies when I finally arrived back in the office, all sweaty, carrying my computer bag and jacket. I finally managed to get a landline call to home and my anxious family. There were no trains out of London that night and so I stayed over.

The next morning, I was on a shuttle bus to a railway station outside London and I got chatting with the lady next to me. She told me how she had been on a tube train behind the one that was bombed on the Piccadilly Line and when they were trapped for hours, she had to deal with a guy that was having a panic attack. She was trying to calm him down when he told her that she didn't understand that he was destined to die soon. It turns out that this guy was late for a Pan Am flight and argued at the check-in desk to be let on—that night, the plane exploded over Lockerbie in Scotland. Then, he was due to have breakfast in one of the twin towers in New York with his brother, who postponed the meeting. So, from his hotel, he saw the twin towers collapse after being hit by airliners with his brother inside. My first thought on hearing all this was that the guy should do the National Lottery as he clearly has lady luck smiling on him!

I was asked a few days later by the CEO to write a review report on how EC Harris management had dealt with the fact that a bus had just been blown up 100 yards from our office in Tavistock Square. The research and my findings from doing that work got me interested in crisis management. It's funny how events shape lives, as my PA at the time, Stacey Smith, had just been trained up as our office first aider expecting to deal with paper cuts and migraines. Instead, she climbed into that smouldering bus minutes after it had exploded and tried to save lives—she went on to become a paramedic.

I worked very long days to get things for a Steelhenge office in Abu Dhabi set up and we got some government clients, but it was such hard work to do on my own.

One time, this created a problem for HSBC when I tried to return a cheque that had been issued by them against our account as a 'Tender Bond', which is required when you bid for major projects. If you win the work but then pull out, the client keeps the bond money to help pay to tender again. I took the cheque into the main HSBC branch in Dubai and the Indian guy behind the counter said they needed a letter to state I was authorised to return it. I explained that I lived 80 miles away in Al Ain and that I was the only representative of the company in the country. No good, he still wanted a letter. I asked for a sheet of A4 paper and handwrote a letter stating that Philip James was authorised by the Regional Director of Steelhenge—to authorise Philip James to return the cheque, and signed it 'Yours Faithfully, Philip James'.

The Indian bank guy was not happy and said the content was fine, but that it needed to be a printed letter, on headed paper and with a company stamp. I told him to f*ck off and pushed the cheque through the counter opening in the glass panel and walked out. The money was back in our account two days later, so I guess they got over having their own cheque back without the formal letter. I tell this story because it is typical of trying to do business in the Middle East in which business operations are often ran by process-loving Indians, probably made that way by the British Empire back in the day.

Steelhenge consultants would fly in from the UK and I would arrange accommodation for them in Dubai so that they could work on the projects. I learned an awful lot about business continuity and crisis management and our sales success improved over time. Unfortunately, our biggest client was an electricity and water company, and they were not happy with what we were trying to implement, and the Steelhenge consultants were too dogged in what they wanted to deliver. A stalemate occurred and the company refused to pay the half a million pounds that they owed us, which, for a small company like Steelhenge, was crippling. I was exhausted dealing with it all and I felt that the owner was about to pull the plug on the UAE operations, when Aziz Zerban, now a regional HSE director at a huge US consultancy called CH2M Hill, called me with an offer I could not refuse, and so I decided to leave Steelhenge.

So, now, in August 2012, I moved to Doha as the Programme Management Consultant CH2M Hill Assurance Director, which was responsible for the Health and Safety, Environment and Security Director for the FIFA 2022 Qatar World Cup. It was an amazing and exciting role working with some of the brightest people I had ever met on an enormous project to build stadiums, training facilities and fan zones. My role included liaison with all the other Qatar government agencies building roads, an entire metro system and a new airport. I had a great team and I was loving the job, but I could see that worker welfare was going to be an issue, and boy, did it become an issue. I wrote strategy papers on what was needed and even developed a plan with a costed one million US dollar semi-permanent workers' accommodation plan that could initially house the workers and then be used for venue staff and even affordable fan accommodation. But the main thing I did was warn them of what would happen if there were construction deaths. Frankly, the client did not care one bit about worker welfare, until CNN, BBC, The Guardian newspaper and New York Times started producing highly damaging articles and undercover reports. Then they were interested, but still not willing to pay for the solutions needed.

I spoke at a conference about the number of deaths that statistically could occur and how we were going to avoid that. I was misquoted in the Qatar press and suddenly I was in trouble with the client—again! In a way, I was glad, as it was obvious to me that the 2022 Supreme Committee were not going to listen to me about worker welfare and in the end, I was glad I was getting away from the whole project.

I offered to resign but CH2M Hill didn't want to lose me, and they said I could work on any of their projects in the world that was not in Qatar. They really pushed for me to go to their Crossrail project in London, but I knew I needed the Middle East salary to recover from my divorce, which had pretty much taken everything I had, and I was still paying a monthly payment to Sam for five years.

In the end, CH2M invented a Middle East regional role for me and I came back to Abu Dhabi. Poor Jules had only just arrived in Qatar from the UAE after we had been apart for five months and we would have another eight months apart whilst she completed her teaching contract at Qatar University. Eventually, we were both back in Abu Dhabi and Jules got another job as Assistant Professor at Rabdan Academy, which trained the military and police. We decided to buy a villa near Abu Dhabi airport and use our rent allowances to pay a mortgage rather than rent; unfortunately, my form with property investment continued and we did this at the peak of the market and even now, nine years later, we recently sold the villa for less than we paid for it, but having rented it for a few years we at least just about broke even.

My new role was very air-travel intensive and after I had spent four months in Saudi setting up a university rebuild project, I was based back in Abu Dhabi. However, I would be visiting projects in Saudi pretty much every other week along with other trips to Oman, Kuwait and one scary trip to Iraq as well as having to manage multiple projects in Abu Dhabi and Dubai.

I did this for three years and was so bored of travelling on planes to visit construction sites that rarely improved, as the clients were just never interested in HSE or enforced any of the standards needed. It was a stressful, tiring and thankless existence. One day, I collapsed at my desk in Abu Dhabi with a suspected heart attack and I was furious as I thought I was about to end my days in such a meaningless way sat at my work desk! During the four days that I was in hospital, I had every test available and was seen by various experts who could find nothing wrong—the opposite in fact, I had the heart of a younger man in terms of health and my veins were clear. They concluded that I may be depressed, and though I thought that ridiculous at the time, on reflection, perhaps they were right.

Anyway, the enforced break meant that I had time to think, and I decided that I needed to change my work situation, and so the next job head-hunter that called me, I would meet for a coffee.

A few weeks later, one did and six months after that, I started, what after my RAF career, was probably the best job I have ever had when I became Associate Director, HSE, of the government owned but privately run Miral, which was developing all the theme parks and attractions on Yas Island in Abu Dhabi, as well as having to develop and implement all the HSE systems for Miral's projects and assets including Yas Marina. My boss was a local construction world legend, Steve Worrell, who, despite being in his seventies, was running about ten projects simultaneously.

In the UK, the word project usually means a new building, a bridge or a highway; but in the UAE, it means building an entire theme park from scratch, such as Warner Bros, SeaWorld, new rides in Ferrari World, two new huge hotels, as well as all the waterfront amenities, pathways and road network. In addition, Miral used to get asked by the Sheikh to do urgent 'little' projects like refurbish the visitor area of the Presidential Palace, build a war memorial visitors centre or a new beach front complex, in record time.

Including a new staff accommodation complex we were building, I had eight major projects on the go with over 12,000 workers, which was a massive undertaking. I was the client representative and so, this time, I was not alone. As the client HSE guy, I was supervising consultants that we used to supervise the contractors building the projects, so I was checking the checkers. Again, I had to develop all the corporate HSE policies and procedures for Miral, but the big difference from Qatar was that Miral did care about the workers on its projects and I had top-level support in this from our Emirati CEO.

I developed worker welfare policies that set new standards in the region and tough HSE prequalification requirements so that the companies working on our sites could only employ reputable sub-contractors. After a few battles with the contractors and companies that supplied plant and equipment, it soon became known that if it was a Miral project, then to only send new and good gear, or they would be replacing it after the first day.

We even had brand-new worker's buses with air conditioning and individual seat belts. You could tell a bus was working on a Miral project as it was shiny, clean and the windows were all closed as the air conditioning worked and the workers were strapped into their seats. Sadly, one bus got forced off the road going back from a site to their accommodation camp and the bus rolled four or five times. Four workers were killed and three more badly injured as they had not been wearing the seat belts provided; the other forty-three workers all survived, badly shaken up but unharmed.

We never had a fatality or serious accident on my watch and I put that down to having good systems in place, good supervision from the consultants, buy-in from the contractors who educated the workers, and a very big dollop of good fortune. On worker welfare, we made sure that our guys were in relatively decent accommodation, which was inspected quarterly. Our managers would turn up at the various worker canteens to have lunch with the workers, which kept the cooks on their toes, and we audited companies to make sure they were paying their guys on time, paying for leave flights and not holding their passports unless the workers requested it.

We also did anonymous worker surveys quarterly and I would stop and question workers with an interpreter during my weekly site walks. We even got the CEO to close Ferrari World early one night so that 5,000 workers could attend and have a nice meal—at first, they were just walking about as they did, not think that they could go on the rides. When we told them they could, they went for it and had a great time. I was told that there was a massive spike in telecom traffic between Abu Dhabi and Asia that evening as thousands of photos and videos were sent by the workers to their families back home. We also used the workers as a 'test crowd' before the opening of the Warner Bros theme park and those workers had a great time too. I am not a 'turn and grind, business as usual' guy.

I like setting projects up or turning around failing ones, and after three years, I was starting to get bored at Miral as everything was ticking over just fine. We had also made enough money by now to think about going back to the UK and retiring, and when Jules' contract ended, we decided it was time to go and I put my notice in. I am very grateful to the UAE for what it has given me in terms of financial reward and more importantly—Jules. But being an expat here is not easy, especially in business, and working knowing that if you upset the wrong person, you can be fired on a whim and so lose not just your job, but your residency and home. The rules are now changing in the UAE to improve things in this area, but change in the region is usually quite slow. There have been some great times though, such as overnight camping in the empty quarter of the UAE desert, weekend dhow trips off Oman, and of course, many of the infamous Dubai weekend brunches. But if, like I did, you spend time most weeks checking an Excel spreadsheet to calculate when you have enough money to leave and have a countdown counter app on your phone to a planned exit date, then it really is time to go!

Then Covid-19 came and our plans to travel around Europe for two years in a motorhome were scuppered. We managed nine weeks but then we risked getting trapped in Italy and came home. Jules found a lovely little Georgian flat in Ripon for rent and so we moved in there. Had Sam still lived in Ripon, I would not have gone back as I didn't think it would have been nice for either Sam or Jules to keep crossing paths, but Sam had gone down south with her new partner and so Ripon was okay for me. I loved being back in the UK, but Jules was not ready for the quiet life just yet and so we had another two-year stint in Al Ain where Jules worked at the UAE University and I stayed retired, and basically just played golf, Football Manager 21 and wrote this book. And guess what, in sleepy little Al Ain, I came across two ex-armourers called Mark Grant and Darren

'Shally' Shallicker, who encouraged my re-entry in the armament world. It seems that armourers are like the Scots and 'manage to get where water won't!'

Jules and I loved the flat in Ripon so much that we bought it, and now both of us call North Yorkshire our home. Jules still gets annoyed when I perfectly reasonably call Yorkshire 'God's country', but she does concede that the weather on the right side of the Pennines is far better than her home city of Salford. It's a pity the football teams are not likewise.

In the twenty years since I left the RAF, the type of work that I have enjoyed the most has been when setting up new major projects or dealing with ones that are in crisis and turning them around. If I could choose again, I would have trained to become a construction project manager as that would have suited my personality and problem-solving skills far better.

The other thing I did, which I really regret now, was that I pretty much shut the door on my RAF past, which meant no reunions, staying as a retired member of the sergeant's mess at RAF Leeming to attend dining in nights or attending any ex-military events; and I didn't even use Facebook when it came out a few years later. Although, I am still not a fan of social media generally, as I think things like Twitter probably do more harm than good, and social media gives a mouthpiece to people that probably don't deserve it. It was writing this book that made me get on Facebook as a research tool and I was amazed at the number of RAF-related groups and the ones specifically for armourers showed me a community of past and some present armourers that really care and try and look after each other. I feel that I have missed out on over twenty years of comradeship, and I feel a bit guilty being a 'Johnny come lately' and that maybe I'm an imposter! Even so, I attended an armourer's reunion last autumn I was part of the motley crew of armourers trying to stay in step as we marched past the Cenotaph on Remembrance Sunday.

So, in conclusion, the twenty years I worked after leaving the RAF were successful for me. I had interesting jobs in construction, travelled extensively for business and pleasure and enjoyed my work most of the time. I also got a professional status way above what I achieved in the RAF, and yet, when I think about when I was happiest and the most content at work—well, that was when I had a blue uniform on and was singing armourer songs!

14. Glossary of RAF terms

ABC – First Aid checks of Airway, Breathing & Circulation
ALM – Air Load Master
AOSC – Aircrew and Officer Selection Centre
AP – Air Publication
APC – Air Practice Camp
ATC – Air Training Corps
AWM – Air Mechanic (Weapons) or Armourer
BFBS – British Forces Broadcasting Service
BSAC – British Sub Aqua Club
BSc (Hons) – Batchelor of Science Degree
CBLS – Carrier Bomb Light Stores
CBU – Cluster Bomb Unit
CO – Commanding Officer
CS Gas – Tear gas
CSE – Certificate of Secondary Education
DI – Drill Instructor
DP – Drill Practice Rifle
DPM – Disruptive Pattern Material
ECM – Electronic Counter Measure
EOD – Explosive Ordinance Disposal
ESA – Explosive Storage Area
FLM – Flight Line Mechanic
GDT – Ground Defence Training
GIT – Ground Instructional Techniques Course
GST – General Service Training
H&S – Health and Safety

HAS – Hardened Aircraft Shelter
HSE – Health, Safety & Environment
IED – Improvised Explosive Device
IRA – Irish Republican Army
JENGO – Junior Engineering Officer
JT – Junior Technician
KR – King's Regulations
LAC – Leading Aircraftsman
LOA – Local Overseas Allowance money
LOX – Liquid Oxygen
MEF – Mechanical Engineering Flight
MFBF – 960 Multi-Function Bomb Fuse
MOB – Main Operating Base
MT – Mechanical Transport
NAPS – Anti Nerve Agent tablets
NATO – North Atlantic Treaty Organisation
NBC – Nuclear, Biological and Chemical
NCO – Non-Commissioned Officer
NVQ – National Vocation Qualifications
OJT – On the Job Training
OTR – Operational Turn Round
PMC – Personnel Management Centre
PSP – Perforated Steel Planking
PTI – Physical Training Instructor
PTSD – Post Traumatic Stress Disorder
QR – Queen's Regulations
QRA – Quick Reaction Alert
R&R – Rest and Recouperation
RAF – Royal Air Force
RFC – Royal Flying Corps
RTU – Return to Unit
SAC – Leading Aircraftsman
SENGO – Senior Engineering Officer
SLR – 7.62mm Self-Loading Rifle
SNCO – Senior Non-Commissioned Officer
SRO's – Station Routine Orders

SS – Selective Service (German Army WWII)
SSO's – Station Standing Orders
SWO – Station Warrant Officer
TACAN – Tactical Air Navigation System
TACEVAL – Tactical Evaluation
TIALD – Thermal Imaging Airborne Laser Designator
UAE – United Arab Emirates
UHF – Ultra High Frequency
WEF – Weapons Engineering Flight
WO – Warrant Officer
WRAF – Women's Royal Air Force
WTC – Weapon Training Cell

Printed in Great Britain
by Amazon